T0324276

Political Biology

Political Biology

Science and Social Values in Human Heredity from Eugenics to Epigenetics

Maurizio Meloni
Department of Sociological Studies, University of Sheffield, UK

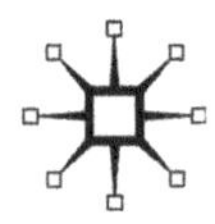

First published 2016 by
PALGRAVE MACMILLAN

Palgrave Macmillan in the UK is an imprint of Macmillan Publishers Limited, registered in England, company number 785998, of Houndmills, Basingstoke, Hampshire RG21 6XS.

Palgrave Macmillan in the US is a division of Nature America, Inc., One New York Plaza, Suite 4500 New York, NY 10004–1562.

Palgrave Macmillan is the global academic imprint of the above companies and has companies and representatives throughout the world.

Hardback ISBN: 978–1–137–37771–5
E-PUB ISBN: 978–1–137–37773–9
E-PDF ISBN: 978–1–137–37772–2
DOI: 10.1057/9781137377722

Distribution in the UK, Europe and the rest of the world is by Palgrave Macmillan®, a division of Macmillan Publishers Limited, registered in England, company number 785998, of Houndmills, Basingstoke, Hampshire RG21 6XS.

Library of Congress Cataloging-in-Publication Data is available from the Library of Congress

A catalog record for this book is available from the Library of Congress

A catalogue record for the book is available from the British Library

To Rebecca, Eva and Marika

The transfer of knowledge and of judgment from one field to another is notoriously difficult, and one need not look far to find men eminent in one field who have made themselves ridiculous by posing as oracles in another. The biologist as sociologist, still more as political prophet or propagandist, runs a similar risk, but we are all necessarily concerned with social evolution. Whether or not they are really pertinent, biological theories are being used in this field, and the biologist necessarily has a part in the discussion, if only as critic.

—G. G. Simpson, 1941

Contents

Preface and Acknowledgments – Problematizing Our Epigenetic Present

There is a palpable excitement around epigenetics. The argument that we must now look beyond our DNA and toward our environment to get a sense of who we are, understand disease trajectories, and perhaps find a way not only to prevent risk but even to improve ourselves is increasingly driving laboratory work and becoming a cliché in the popular media. Soon policymakers and health experts will likely be on board as well. And the high priests of DNA-centrism are rapidly reconfiguring their language to meet the challenge of epigenetics.

To wit, Francis Collins, director of the National Institutes of Health (NIH) and the dean of the Human Genome Project, told a conference in San Francisco that the expression "junk DNA" is a sign of hubris from the past: "Most of the genome that we used to think was there for spacer turns out to be doing stuff and most of that stuff is about regulation and that's where the epigenome gets involved, and is teaching us a lot" (Collins, 2015). "In human diseases," the journal *Nature* advises, "the genome and epigenome operate together. Tackling disease using information on the genome alone has been like trying to work with one hand tied behind the back. The new trove of epigenomic data frees the other hand. It will not provide all the answers. But it could help researchers decide which questions to ask" (*Nature* Editorial, 2015). In February 2015, the journal introduced readers to the Epigenomic Roadmap (Skipper et al., 2015). And, as the Federation of American Societies for Experimental Biologies accurately and succinctly puts it, "While the discovery of the genetic code led researchers to believe that our physical appearance and susceptibility to certain diseases were 'hard-wired' within our DNA, exciting advances in our understanding of the human genome have shown that this is not the entire story. Scientists now know that both biological and environmental factors play an important role in how we develop and age and even in determining our risk of diseases like cancer, cardiovascular disease, and type 2 diabetes" (FASEB, 2014). Finally, NIH is now funding investigation on the link between paternal nutrition and offspring metabolism.

In anticipation of an avalanche of articles, books, hype, and controversy – in which epigenetics will be mobilized to 'explain' sexual orientation or the transgenerational effects of various historical catastrophes – this book offers its own roadmap to our epigenetic present. It tells the story of how the biological notion of heredity became modern by discarding epigenetic-like hypotheses and constructing, in the late nineteenth century, what is known as hard heredity: the idea that there can be no environmental influence on hereditary material. The twentieth century was increasingly dominated by hard-hereditarian explanations, with a range of implications – progressive and regressive, inclusionary and exclusionary – for politics, public policy, social values, and the boundary between biology and the social sciences. But in recent decades, the hegemony of hard heredity has eroded. We are again thinking about heredity in an extended way by incorporating ancestral behavior and experience into our stories of generational transfer.

What sort of change can we expect from this new dynamic? What sort of citizenship, personhood, politics, and governmentality will emerge from this porous view of the biological body, shaped by today's experiences and those of past generations? In the pages that follow, I rely on history to show the broken logic behind the assertion that this alternative notion of heredity will *necessarily* have better social policy implications than has staunch DNA-centrism. To grasp meaning in our present, we need what many scientists and social scientists working on these topics like to avoid: history, history, and again history, the neglected story of how we came to think in certain ways and rule out others. If this book even partly succeeds in its archeological mission to excavate and problematize the sources of the present, its goal is achieved.

I am grateful to many people who have helped me to think through the research and arguments of this book and to find a path through the nuanced debates on human heredity and its implications for policy, social values, and knowledge production.

In the United Kingdom, I have benefited from exciting exchanges with Peter Bowler (Belfast), John Dupré and Staffan Müller-Wille (Exeter), Des Fitzgerald (Cardiff), Simon Williams (Warwick), Paul Martin and Vincent Cunliffe (Sheffield), Martyn Pickersgill (Edinburgh), Brigitte Nerlich and Aleksandra Stelmach (Nottingham), Chris Renwick (York). In the United States, I had the huge privilege of writing this manuscript in the course of a yearlong membership at the Institute for Advanced Study, School of Social Science, Princeton, NJ. My deep thanks to the Institute for offering me this opportunity. While there, I benefited from profound intellectual discussions with Danielle Allen (now at Harvard),

Didier Fassin, Joan Scott, and Michael Walzer and with my co-members during the academic year 2014–2015, particularly Jennifer Morgan (NYU), Michael Hanchard (Johns Hopkins), Hugh Gusterson (George Washington), Gary Fine (Northwestern), Matthew (Brady) Brower (Weber State University), Nicole Reinhardt (Durham), John Holmwood (Nottingham), and Gurminder Bhambra (Warwick), Brian Connolly (South Florida), Serguei A. Oushakine (Princeton), K. Steven Vincent (North Carolina). I thank the library staff at the IAS for their assistance and patience with my serial requests: Kirstie Venanzi, Nancy Kriegner, Karen Downing, Cecilia Palmar, all of whom aimed in this book's development. A big thanks also to Donne Petito and the other staff members at the IAS School of Social Science.

Michael Gordin (Princeton) offered kind advice and comment on Chapter 4. He also put me in contact with Loren Graham. My meetings with him, Evelyn Fox Keller, Diane Paul, and Gar Allen were memorable. Betty Smocovitis (University of Florida) was incredibly helpful in framing the role of the modern synthesis in the politics of biology. I had also extensive and very stimulating correspondence with Mark B. Adams (University of Pennsylvania). My thanks to Stephanie Lloyd (Laval) and Eugene Raikhel (Chicago), who organized a workshop on epigenetics where I presented what became Chapter 4. I thank profoundly Margaret Lock (McGill), Jörg Niewöhner (Berlin), Sarah Richardson (Harvard), Giuseppe Testa (Milan), Sahra Gibbon (UCL), Liz Roberts (Michigan), Rayna Rapp (NYU), and Hannah Landecker (UCLA) for their insightful comments. During the preparation of the manuscript, I had the pleasure of corresponding with Eva Jablonka (Tel Aviv), always so kind and available to dispel my confusions about epigenetics. Frances Champagne (Columbia) and Tobias Uller (Oxford) provided valuable help with the revision of an article that is now incorporated in Chapter 7. I have also benefited from correspondence and discussion with Paul Griffiths and Karola Stotz (Sydney), who helped me orient myself in the world of postgenomic research. Needless to say, all the possible errors in this book are my own.

As a non-native speaker, I've relied more than a little on copyeditors. David A. Walsh, a graduate student at Princeton, lent an important hand. Andre Turner (Bristol) helped with articles that now form part of Chapter 7. But, above all, Simon Waxman, an independent editor in Boston, has had a key role in making this book readable and possibly even enjoyable, working around the convolutions of my English. Special thanks to Joan Scott for connecting me to Simon.

At Palgrave Macmillan, Holly Tyler and Dominic Walker have been extremely kind and supportive throughout the process.

I gratefully recognize the Institute for Advanced Study and its director Robbert Dijkgraaf for providing funding. The research behind this book is also supported by a Leverhulme Trust grant in the Department of Sociological Studies (Principal Investigator Paul Martin) at the University of Sheffield, where I am currently based.

Finally, this book is dedicated to my three ladies, two of whom came into the world while I was writing it. I owe incalculable debts to Marika. I hope that Rebecca and Eva, the new pair, will one day look at their father's book and find it still a meaningful guide to the intellectual controversies driven by the scientific advances of their time.

References

Collins, F. (2015). Speech at the JP Morgan Healthcare Conference on January 13, 2015. Available at: http://sandwalk.blogspot.ca/2015/01/francis-collins-rejects-junk-dna.html. Accessed October 5, 2015.

FASEB (Federation of American Societies for Experimental Biology). (2014) "Looking beyond our DNA". Available at: https://www.faseb.org/Portals/2/PDFs/opa/2014/Epigenetics%20Horizons.pdf. Accessed October 5, 2015.

Nature Editorial (2015). Beyond the genome, 518, 273 (February 19, 2015). doi: 10.1038/518273a.

Skipper, M., Eccleston, A., Gray, N., Heemels, T., Le Bot, N., Marte, B. et al. (2015). Presenting the epigenome roadmap. *Nature*, 518, pp. 313–314.

1
Political Biology and the Politics of Epistemology

The historical context: the long twentieth century of the gene and beyond

This book is an exploration of the diverse transformations that have occurred in the space between biology and politics during the twentieth century, when the modern view of heredity, epitomized by genetics, took shape, was consolidated, and finally was challenged again.

Although the rediscovery of Gregor Mendel and his work bears the felicitous date of 1900, so that the last century really is definable as "the century of the gene" (Keller, 2000), one must look beyond both its beginning and end to assess thoroughly the social significance of the genetic view of heredity. It was, in this sense, a long twentieth century.

First, one must go back farther to understand the epistemic conditions that made Mendel's rediscovery intelligible at all (Bowler, 1989). I refer in particular to the elaboration of hard heredity by Francis Galton and August Weismann in the last decades of the nineteenth century, an effort that would be completed by Danish botanist Wilhelm Johannsen in the early twentieth century. Hard heredity, the notion that the hereditary material is fixed once and for all at conception and unaffected by changes in the environment or phenotype of the parents (Bonduriansky, 2012), can be seen as the key conceptual move that created the epistemic space within which the Mendelian notion of a particulate and stable hereditary material (later christened the gene) could be situated.

The move toward hard heredity was a radical break with the popular and scientific views of heredity that dominated the late eighteenth and nineteenth century, especially in medical writings.

These views, which we can call for simplicity *soft hereditarian* (actually a much later and somehow questionable terminology)[1], converged on the idea of a direct and formative influence of the environment on the hereditary material. Heredity in the soft version is affected by the parents' or grandparents' lifetime experiences, not fixed at conception (Bonduriansky, 2012). The hereditary material can be modified "either by direct induction by the environment, or by use and disuse, or by an intrinsic failure of constancy." The modified genetic material would then be transmittable to offspring (Mayr, 1980: 4).

The widespread belief in the inheritance of acquired characteristics (Zirkle, 1946) was historically identified with Lamarckism (though largely preceded it) or, after 1880, Neo-Lamarckism. In 1909, two French biologists, Yves Delage and Marie Goldsmith put it this way:

> Whatever theory emphasizes the influence of the environment and the direct adaptation of individuals to their environment, whatever theory gives to actual factors the precedence over predetermination can be designated as Lamarckism. (translation, 1912)

Much could be passed from parents to offspring before hard heredity became the accepted view. For instance, according to the theory of telegony, even the characteristics of a previous mate were believed transmittable from a mother to her offspring. Some animal breeders held fast to this model as late as the early twentieth century.

After 1900 hard heredity crystallized as the main and, eventually, only possible view of heredity. This was truly a revolution, setting in motion vast changes in the general understanding of the relationship between human beings and their biological substance. But its impact went beyond even this, reaching also into politics and the organization of knowledge. As we will see, hard heredity fostered new demarcations between the ontological domains of the biological and the social, nature and nurture, the life and the social sciences.

Although hard heredity was ascendant for decades, its domination has lately been challenged. The late twentieth and early twenty-first centuries have brought major conceptual challenges to the once-firm concept of the gene. These challenges are gathered today under the broad umbrella of the *postgenomic* age. Today scientists are exploring a number of gaps in our knowledge of genes, unforeseen complexities surrounding the hereditary material. These studies have revised the view of the genome in a seemingly backward direction: no longer fixed from birth but instead deeply affected by environmental signals, from cell to society.

The genome of the twenty-first appears to be, as Barbara McClintock anticipated, "a highly sensitive organ" (1984; see Keller, 1983). It has become a "developing genome" (Moore, 2015), subject to time and space, biography and milieu (Lappé and Landecker, 2015). A "reactive genome" (Gilbert, 2003; Keller, 2011, 2014; Griffiths and Stotz, 2013) or a "postgenomic genome" is "an exquisitely sensitive reaction (or response) mechanism" (Keller, 2015), powerfully influenced by all sorts of biological, but also "sociocultural" pressures, originating in the individual body or society at large: toxins, work stress, nutrition, socio-economic status, maternal care, and grandparental lifestyle. This new understanding is a paradoxical product of scientific advances that were expected to deepen and confirm preexisting theories of the fixed gene, but as Evelyn Fox Keller noted even fifteen years ago, "contrary to all expectations, instead of lending support to the familiar notions of genetic determinism" advances in genetics have posed "critical challenges to such notions" (2000: 5).

One has simply to look at the huge explosion of the literature on epigenetics to see how entrenched this reactive view of the gene has become in a relatively short time. Epigenetics concerns environmental regulation of gene expression. Rather than just a present mania, it would be better to understand epigenetics as the last chapter of an often neglected but very honorable story of going "beyond the gene," (Sapp, 1987) looking at various "layers of non-Mendelian inheritance" (Richards, 2006; Jablonka and Lamb, 2014). Although epigenetics is often employed in a broad and nebulous sense – and its capacity to infiltrate language, and therefore its success, partly depends on this vagueness (Meloni and Testa, 2014) – a significant part of research on epigenetic mechanisms implies possible transgenerational effects (Jablonka and Raz, 2009). Thus biologists are suddenly discussing again "soft inheritance" (Richards, 2006; Gilbert and Epel, 2009; Bonduriansky, 2012) or "inheritance of acquired characters" (Smythies, Edelstein and Ramachandran, 2014) or even the "dirty word," as Ernst Mayr called it, of Lamarckism (Jablonka and Lamb, 1995, 2005, 2008; Vargas, 2009; Gissis and Jablonka, 2011; Baiter, 2000; see also Burkhardt, 2013).

This turn to soft heredity has not remained confined to biological debates. An array of sociological, anthropological, and epidemiological studies now focuses on the idea that early-life developmental factors can have an influence not only on adult life but also, potentially, on the next generation(s). Claims that experiences of past generations can be transferred to later generations via epigenetic processes and influence, for instance, individuals' vulnerability to disease, are growing at a rapid pace. Ancestral obesity or malnutrition, prepubescent paternal smoking, the shock of 9/11 and the Holocaust, the 1918 pandemic influenza, and

American slavery have all been considered as contemporary sources of ill health or the cause of some sort of present epigenetic difference, if not abnormality.[2] "Soft inheritance has now been reborn," say Mark Hanson and Peter Gluckman (Hanson, Low, and Gluckman, 2011), leading promoters of the idea of *in utero* programming of adult vulnerability to disease.

It is important to clarify, however, that epigenetics and the wider notion of the reactive genome does not return us to soft heredity in the sense that prevailed before Weismann (although interestingly there are claims that even some apparently bizarre ideas such as telegony have been "rediscovered", see Crean, Kopps, and Bonduriansky, 2014). History is a not a pendulum that swings back and forth between otherwise immutable positions. It makes more sense to speak of a transformation of Lamarckism (Gissis and Jablonka, 2011) or a modification of the soft vs. hard heredity debate (Moore, 2015). What circulates among today's researchers and authors is an appeal for a richer and more "pluralistic model of heredity" capable of combining "genetic and nongenetic mechanisms of inheritance" and of recognizing "the reality of both hard and soft inheritance, and the potential for a range of intermediate phenomena" (Bonduriansky, 2012).

Nonetheless, it cannot be underestimated how traumatic this view can be for the sort of progressive story told by the fathers of the modern evolutionary synthesis (Mayr, 1982), in which a one-way march from soft to hard heredity (from darkness to light!) was the only possible direction of history.

This historical arc, from Weismann to epigenetics, from the making to the unmaking of hard heredity, from the narrowing of heredity to its present broadening, from its modernist reification (Müller-Wille and Rheinberger, 2012) and purification to postmodernist dispersion and complexification, in sum the transition from Heredity 1.0 to Heredity 2.0 (Meloni, 2015a) is the intellectual background and conceptual battlefield where I look at the political implications of human heredity.

Biology and social values

There are many fruitful directions from which to consider the connection between biology and social values. One can engage contests over human origins, the politics of sex and gender in biology, the doctrine of the cell state, animal social behavior, and the use of immunological and organic metaphors in human society. However, investigating the connection between biology and social values from the viewpoint of soft-versus-hard heredity, as I do, is a particularly productive choice for several reasons.

First, as I have noted, *this is a timely choice* given the new influence of epigenetics and the controversial evolutionary question that arises thereby: whether this openness of the genome to external influences indicates a return to rejected views of soft heredity. Thanks to the present ascendancy of epigenetics, we are now able to look at the story of hard heredity that started with Galton and Weismann genealogically, in all its precariousness and even finitude. Without the present destabilization of hard heredity, the founding events of the modern view of biological inheritance would seem more fixed and taken-for-granted, less open to an "excavation" (Daston, 2004) that consigns these events to their contingency. We are in a position to wake them from what Michel Foucault called "silent monuments" (Foucault, 1972).

This book is therefore an *archaeology of the past* – going back to the making and consolidation of hard heredity and exploring its epistemic and political implications for the social sciences and wider society – made possible by *a tension in the present,* namely current challenges to hard heredity and their implications for evolutionary theory, the social sciences, and society at large. And this archaeology allows claims about a *possible future* in which the rift between biology and society – which marked the twentieth century as a consequence of a certain view of heredity, the gene, and biology in general – is bridged.

Second, I choose this soft-versus-hard debate because it is a good point of entry point into wider contests over *epistemic values* and knowledge production, in particular the barriers between the life and social sciences. Whether the experiences of one generation pass on to the next is a conceptual and logical watershed in defining the boundary between the biological and the social. In a soft-heredity view, the social is always on the brink of becoming biological, habits are turned into instincts, and life experiences of a previous generation are embedded in the biology of a successive one. Lamarckism or soft heredity is the condition for a fully biosocial or biohistorical investigation, for a continuous exchange of the biological and the social. Not by chance, when Lamarckism was the dominant view amongst social scientists, from Spencer to early 1900, the autonomy of the sociocultural was just a mirage.

By contrast, hard heredity maintains separation between the social and the biological. Genes are passed on, but the lifetime experience of each generation is canceled in the next, and each generation has the chance to start anew, as Weismann noticed. No significant sociocultural experience can leave a mark upon the hereditary materials (not necessarily bad news, as Alfred Russel Wallace noticed, considering how often significant life experiences have been for human society negative and even catastrophic ones, from famine to war). None less than the pioneering anthropologist Alfred Kroeber embraced hard heredity to build an idea of the sociocultural as super-organic and independent of biological influences (Kroeber, 1915, 1916a, 1916b, 1917; see Kronfeldner, 2009; Meloni, 2016). Twentieth-century anthropology, like twentieth-century evolutionary thought, had to cut the Lamarckian knot – to be rid of the "confusion" between the social and the biological in order to establish clear epistemic boundaries between disciplines.

Finally, few debates have been so politicized in the history of biology as the application of soft and hard heredity to human affairs. I have always considered it very telling that in 1928 Anatoly Lunacharsky, the first Soviet commissar for education, decided to screen the tragic story of Austrian zoologist Paul Kammerer, whose suicide, two years before, marked in a sense the end of the "golden age" of Lamarckism (Gliboff, 2011), at least in Europe. Lunacharsky saw in Kammerer's story a source of politicization. He and some fellow Bolsheviks understood the struggle for the scientific legitimacy of the inheritance of acquired features as a contribution to a more progressive, socialist view. However, not everyone in the Soviet Union agreed with Lunacharsky's ideas about the meaning of soft inheritance for the socialist revolution, leading to intense debate that I will discuss in the chapters that follow. For this other group, people like Yuri Filipchenko, soft inheritance just meant that social inequality could be passed on from society to the genes of poor people, who were then doubly disadvantaged: socially and biologically. In some respects, this debate is still with us, now that the ascendancy of epigenetics has restored soft inheritance to a place of concern in the realms of public policy and public health.

An exercise in political epistemology: setting up a research program

Conjoining what is disjoined

This book explores the entanglement of two apparently distinct subjects: first, the political translation into the public sphere of certain debates in biology (such as hard versus soft heredity) and their polysemic association with sociopolitical values; second, the reconfiguration of the border between life and social sciences following the rise and fall of certain scientific views of human heredity.

Prima facie, the two problems seem distant from each other. No matter the efforts of various traditions, from Foucauldians to science and technology scholars, to write a conjoined history of politics, science, and the social sciences (Jasanoff, 2004), the academic departmentalization of our intellectual life has proved so far sufficiently leak-proof to situate the first theme under the label of "politics" and the second as a matter of history or sociology of science. However, real history doesn't follow such boundaries. One of the arguments of this book is precisely that, in the history of human heredity, there exists a dangerous and inextricable, irreducible affinity between epistemic[3] and political factors, the organization of knowledge and political events. This entwinement between knowledge and politics has been undertheorized so far, but it is at the core of the events I describe. For instance, Galton, as everyone repeats, was the founder of both the nature/nurture dichotomy (ordering of knowledge as a consequence of his politics) and the eugenics movement (ordering of society as a consequence of his scientific view). This coincidence is well-noted, of course, by several scholars (firstly, Cowan, 1972a, b; 1977). However, has the *interplay* between knowledge production and political intervention, between Galton the dichotomous ideologue of nature-nurture and Galton the eugenicist been sufficiently appreciated?

It seems to me that scholars, with very few exceptions, have been successful enough in neutralizing this entwinement between "knowledge of heredity" and "formative power" (Müller-Wille and Rheinberger, 2012) that establishes itself at the very beginning of the story narrated in this book. As a consequence of this neutralization, histories of eugenics as a chapter in political science are rarely, if ever, genealogies of the social sciences as a way of organizing a distinctive form of knowledge. Meanwhile, histories of the social sciences as autonomous disciplines rarely look at the broader political context in which social scientists interacted with biologists, eugenicists, and views of heredity. These histories offer little perspective on the questions I ask: Why did neo-Lamarckians recoil in horror at the prospect of hard heredity? Why did Kroeber embrace it enthusiastically? Why, after World War II, did anthropologist Ashley Montagu and geneticist Theodosius Dobzhansky evangelize the uniqueness of man and the irreducibility of culture? What was all the fuss when sociobiologists denied this uniqueness three decades later? Finally, why, with today's postgenomic genome and even claims of a return to soft heredity, the purifying separation between the social and the biological seems to wane and new biosocial or biocultural questions re-emerge on the agenda of sociologists, social theorists and anthropologists (Ingold and Palsson, 2013; Meloni, Williams, and Martin, 2016; Frost, 2016)?

My notion of political biology as a political epistemology of the life sciences is a deliberate hybrid that conveys the inextricable, messy interconnection of epistemic and political events in the relationship (and genealogy) between biology and social facts, and between the life and social sciences. Only keeping in sight this Janus-face perspective on politics *and* the organization of knowledge that many of the points of tension between biology and the social can be better understood and possibly ironed out. However, before exploring more in details what I mean by political biology and how I mean to use it as a heuristic tool, it is necessary to situate my book in the context of recent social science debates.

Political biology in the context of recent social science debates

My project for a political biology, lies at the intersection of three main clusters of problems and lines of thinking: (1) sociology of scientific knowledge (SSK) and science and technology studies (STS), (2) biopolitics, and (3) the project for a historical ontology and epistemology.

Regarding the first area, my project is receptive and indebted to SSK and STS, although it aims to differentiate itself from some shortcomings of the two fields. For instance, I don't feel comfortable with some constructionist excesses from these two quarters that have turned science and its history sometimes into a mere epiphenomenon of the wider social context, in spite of (laudable) claims to keep together in a network nature, society and their collective representation (Latour, 1993). My wish to accommodate a genuine dialectics between scientific knowledge and social values is revealed by my effort to pay the most serious homage to the enormously fascinating history of modern biology and human heredity in particular. One can make room for conflicts and pluralism without turning truth and objectivity into mere fables, and a passion for history is often the only antidote to this drift.

Also, I aim to inject a higher dose of political awareness into STS. The field has brought science down to earth and helped us look at scientists as "socially situated reasoners" (Knorr-Cetina, 1981), thus destabilizing "the special kind of naturalization created by scientific and technical representation" (Jasanoff, 2012). This is in itself a profoundly political gesture. But the politics of STS is very often micropolitics, a politics of technical decisions and of science for liberal-democratic, advanced societies, in which citizens possess "agency with respect to the production and application of scientific knowledge" (Jasanoff, 2005: 266). Much is taken for granted – a right to knowledge, critical capacity, and self-determination – and much else is ruled out or forgotten, such as science's hugely transformative power in the social arena (for good or ill) and scientists' phantasies of power. Life-scientists can imagine utopian and dystopian scenarios, play at dictatorship, or try to persuade dictators to rule on behalf of their own agenda.

This messianic and quasi-totalitarian view of biological science, so prominent until the 1940s in the West and the Soviet Union, is often neglected in contemporary STS. So is the broad humanistic attitude of the architects of the modern biology (the modern synthesis) in post 1940 biology, from Theodosius Dobzhansky to George Gaylord Simpson. People like Muller and Huxley, Dobzhansky and Simpson, were not simply socially situated reasoners, as STS wants, but also socially situated militants, even partisans, and if we want to really "follow scientists around society" (Latour, 1988), we also need to follow them when they participate, directly or indirectly, in the broader ideological debate of their own time, imbuing their own science with social and political values.

It is at this level that biopolitical writings could provide a helpful supplement to the micropolitics of STS. But here, too, dissatisfaction emerges. My political biology is an explicit alternative to biopolitical writings, antagonistic not so much to Foucault's historically oriented (but ultimately very much programmatic) project for a biopolitics, but to the way biopolitics played out in recent traditions of thought, especially amongst social theorists. Here in particular, recent biopolitics has forgotten the passion for history, history of science in particular. I will argue in a next section that a serious biopolitical approach cannot exist without a deep understanding of the history and conceptual issues surrounding the notion of life and the life sciences, with their epistemological dilemmas.

Lastly, the book is indebted to the (largely Foucauldian in inspiration) projects of historical ontology (Hacking, 2002) and historical epistemology (Lecourt, 1969; Daston, 1994; Daston and Galison, 2010; Rheinberger, 1997, 2010). This means that I focus on the epistemic conditions behind the emergence and disappearance of scientific constructs. I therefore also emphasize caesuras, epistemic breaks, dispersions, switches and discontinuities rather than epistemic homogeneity, progress, and cumulative development in history of scientific thought (Müller-Wille and Rheinberger, 2012).

Before coming to the main argument of the book, I will explain in more detail the positioning of my project with regard to these three clusters of problems.

Rethinking the intersection of the natural and the social order

First, this book can be seen as an attempt to go back, once again, to the relationship between the natural and the social order, the body natural and the body politic. This is a concern in much political thought, in SSK (Barnes and Shapin, 1979; Shapin and Schaffer, 1985) and STS (for instance: Latour, 2009; Jasanoff, 2004) all traditions that have tried to explore how "solutions to the problem of knowledge" are often a "solution to the problem of social order" (Shapin and Schaffer, 1985). By focusing on the political and scientific dynamics around human biology, and human heredity in particular, I expand on the debate within a substantial body of scholarship, which, since the 1970s, has looked at that juncture where power and knowledge, epistemology and social influences, science and values are so inextricably interwoven to be, in a contemporary idiom, co-constituted or "co-produced" (Shapin and Schaffer, 1985; Jasanoff, 2004, 2005, 2012).

Such a co-productionist approach has the great advantage of bringing complexity into the science/society debate by escaping both the modernistic autonomization of scientific rationality from social influences and the mere re-inscription of the scientific discourse into an effect of power, avoiding the primacy of one of the two sides. As Jasanoff writes (2004: 16): "Science in the co-productionist framework, is understood as neither a simple reflection of the truth about nature nor an epiphenomenon of social and political interests" (see also Reardon, 2005).

However, the fact that scientific reality and social reality are in a sense two sides of the same coin cannot be taken as an a priori slogan, but has to be investigated in its nuances and contingencies, showing the often dialectical and complex way in which this co-production effectively takes place, for instance in my case, in debates on human heredity, with its many tensions, resistances and discrepancies from both sides of the coin. Of course, the visibility of political influences and ideological values on the framing and production of knowledge in biological sciences are a well-established fact for historians who have looked at cultural settings that are *alien* from standard Western ones. So, for instance, there is an abundant literature on the "symbiotic relationship" (Weiss, 2010) of politics and science in Nazi Germany. An even clearer example is the widespread literature on the Lysenko affair, a sort of textbook case for the distortion of science by political ideology. When things go really wrong, the standard argument on Lysenko goes, science becomes politicized and plays into the hands of tyrants and charlatans.

The idea that science is politicized only, or mostly, in totalitarian regimes and that democracy applies only a minimum pressure to science (therefore speeding its progress) is an established theme in the sociology of science (Barber, 1952; see also Merton, 1938/1973, 1942; Polany, 1962;). Especially in the post-war years, this discourse was part of a broader social ideology, the idea, designed in opposition to Nazism and Lysenkoism and epitomized in the work of Vannevar Bush and the National Science Foundation (Science: the Endless Frontier: Bush, 1945), that knowledge "advances the most quickly and surely when its pursuers are liberated from social influences of any kind" (Hollinger, 1998:102–103; see also Kevles, 1977; Fuller, 2000).

Given this idealized legacy (Jasanoff, 2004) it has been difficult to appreciate the complexities of the relationship between politics and science, especially when this influence: (1) affects the standard history of Western science and not only alien cultures or pseudo-science; (2) concerns the present and not the past; (3) regards "good," democratic ideologies rather than totalitarian ones, agreeable moral goals not horrific ones (see Proctor, 2003).

It may in sum just be a form of ethnocentrism to believe that "externalist forces" are at work "only in 'exotic' locales, and not in 'normal' ones" (Graham, 1993: 202), or in the past and not in the present, or for the bad and not the good (Proctor, 1988, 2003). What we have learnt from SSK and STS is that science is "never pure" and the strength of the boundaries (Barnes, 1974) between the internal and the external (Shapin, 1992, 2010) has to be proved also for our normal science and for our time. We have also learnt that it is part of the process of purification of scientific claims to demarcate themselves from the politicization of science and keep this boundary in function (Gieryn, 1999; Latour, 1993) to establish its authority and legitimacy – like Athena, fully formed out of Zeus's head. In this sense, the politics of science and, in my case, political biology is absent from view exactly because it is what has to be kept conceived in the process of constructing the authority of science.

The external and the internal in the science-society relationship

In saying that externalist forces are at work in our normal Western biology, however, I want to avoid two naïve assumptions.

The first is that the external and internal, the political and the scientific, can be partitioned such that the socially aware investigator must them join them artificially. There is no such partition, and the fusion is not artificial. There is instead a dynamic space where the two orders visibly exchange before they are bifurcated and ossified into two apparently incommensurable regimes of truth.

Bruno Latour calls this division the moment when matters of concern are turned into riskless matters of fact (Latour, 2008). Neither interests nor society are ready-made and easily separable (Knorr-Cetina, 1983). As Sheila Jasanoff writes, taking seriously the social character of scientific knowledge implies that

> Facts and artifacts...do not emerge fully formed out of impersonal worlds, with cultural values entering into the picture only when a technology's impacts are first felt; nor, by the same token, does sociality enter into the making of science and technology as a secondary player, by side doors only. (2005: 269)

It is therefore necessary to think in terms of the entanglement (or "generative entrenchment," Rheinberger, 1997) of science and politics, rather than their after-the-fact interconnection.

This also implies an anti-purist methodology that denies that social factors should be seen as non-rational (Laudan, 1977), non-cognitive, and irrelevant or, worse, "contaminants" of scientific rationality. They are rather "constitutive of the very idea of scientific knowledge," full participants in an enlarged version of what counts as scientific rationality (Shapin, 1995; see Barnes, 1974; Bloor, 1976; Knorr-Cetina & Mulkay, 1983; Longino, 1990; Longino, 1996). What has to be understood and explained is the neutralizing language that transports temporally contingent contexts and politics "outside of science" (Shapin and Schaffer, 1985).

However, one of the assumptions of this book is that the construction of scientific discovery with and upon social facts does not imply a nihilistic view of science. Science can keep its authority not in spite but because of its social construction: all scientific beliefs are socially constructed and politically motivated to a certain extent, but not all are equally valid or invalid because of that. We still can strive for objectivity and truth while recognizing that scientific statements are situated in the social world (Haraway, 1988; Keller, 1985). Looking at histories of human heredity from the situated viewpoint of political biology is not an exercise in truth-dissolution, but a broadening of the history of science to its wider context by using the "privilege of partial perspective" (Haraway, 1988).

The second naïve assumption would be to dissolve scientific knowledge into its contextual factors, thus denying the reciprocal influences, and the two-way traffic between the original epistemic dynamic of science and its sociopolitical setting. Scientific findings are partly the plastic object of power and ideological structures, but they also *resist*, *shape*, and *produce* them. To have a genuine dialectics – based on friction or "signals of resistance" (*Widerstandsaviso*) in Fleck's term (1979) between findings and theories – we need to presuppose that science exceeds the constructed triad of power, interest and social engagement (Kuhn, 2000). Science and knowledge are the plastic objects of power, but so is politics the plastic object of scientific facts and expert claims. Therefore power and truth, matters of fact and matters of concern, loop back on each other.

Without this appreciation of the active, polysemic, and often unpredictable implications of scientific facts and the sociopolitical trajectories they put in motion, this book would be merely a restatement of the superiority of social over objectified knowledge, of society over nature, one of the mantras of Western thought that have been so unproductive in understanding the creative and imaginative power of science in shaping the political landscape at large. I join authors such as Evelyn Fox Keller, Donna Haraway, Karen Barad (2003, 2006, 2008) and Elizabeth Wilson (2012, 2015) who oppose the postmodernist dissolution of reality into ethereal discourse with a materialist ontology of what scientific discourse is really about. In the context of what is known in social theory as "new materialisms" (Coole and Frost, 2010; see also Alaimo and Hekman, 2008; Bennett, 2010), this return to the "weightiness of the world" (Barad, 2003) offers a chance to bring the science wars to a close.

Political biology as a political epistemology of the life sciences

I explore the profound entanglement of knowledge and sociopolitical values in the history of the biology-society relationship in terms of *political biology*. Political biology is the application of political epistemology to the history of biology. But what exactly is political epistemology?

Political epistemology comes in many guises. Bruno Latour, for instance, has repeatedly used this label in his attempt to challenge the "epistemological police" (2004; see also 1987) that patrol the borders of knowledge with reassuring disciplinary divisions between science and power. Jasanoff has proposed not only that science and technology are "*political* agents" (2004) but has also spoken of "civic epistemologies" or epistemology as "culturally specific, historically and politically grounded, public-knowledge ways" (2005: 267). Jasanoff's project looks at the way in which power becomes an epistemological force while knowledge is "generated, disputed, and used to underwrite collective decisions" in a continuous unfolding process (2005: 24; see also 2012).

In political theory and philosophy as well, the notion of a political epistemology is gaining importance. Analytic philosopher Miranda Fricker, has made appeal to "a politicization of epistemology" (1998) as a way to contrast "epistemic injustices" linked to the role the concept of knowledge plays in human life (see also 2007). In *Politics and the Criteria of Truth* (2010), political theorist Alireza Shomali defines political epistemology as "a research project about how politics frames the questions of knowledge... a strategic activity to make the production of truth" and promote "a new understanding of the connection between political engagement and epistemic normativity."

However, the most apt usage of political epistemology for the purposes of my project is undoubtedly Steven Shapin and Simon Schaffer's in *Leviathan and The Air-Pump* (1985). In their classic study, Shapin and Schaffer investigate the political context of the 1660s Hobbes-Boyle debate on the vacuum. Here scientific statements – for instance, claims on behalf of the existence of the vacuum – are read as part of a broader political negotiation between stakeholders. Scientific statements become elements in the "political activity in the state": the rejection of the notion of the vacuum is understood as the "elimination of a space within which dissension could take place" (1985: 109). Shapin and Schaffer argue that the "history of science occupies the same terrain as the history of politics" and that any solution to the problem of knowledge is a political solution. As they write:

> The problem of generating and protecting knowledge is a problem in politics, and, conversely... the problem of political order always involves solutions to the problem of knowledge. (1985: 21)

Although the translation of a seventeenth-century debate to the context of the long twentieth century of genetics can only be a loose one, the idea of using "ontology and epistemology to secure public peace" inspires my attempt to read the history of science at the interface of scientific statements and political order.

What is "politics" in political epistemology?

Before proceeding further to my understanding of political biology I need to say more about what I mean by *politics* in this context.

My usage of politics has three main sources. First, in the tradition of the social study of science, there is a broad Latourian sense in which politics refers to the long chains of associations that epistemic facts mobilize, continuously turning matters of fact into matters of concern, and vice versa (2004, 2008). Latour points at that "muddle" where objects appear as hybrids, before the "modern settlement" might produce any "sharp separation between their own hard kernel and their environment" (2004).

This co-production of scientific facts and political issues can also be alternatively expressed using a second set of terms. I refer to the Foucauldian alternative between analytic of truth and ontology of ourselves (Foucault, 1986: 85–86). The epistemic moment captured by the formal "analytic of truth" approach (the investigation of "the conditions in which a true knowledge is possible") is always on the point of being reversed into a substantial and historically situated "ontology of ourselves" that is a moral and political moment through which human conduct and ethos are constantly reconstituted. The cognitive, analytic moment of science becomes a source of profound politicization, through which human beings are classified and governed and may find sources for their own self-fashioning and even emancipation.

A third influence on my notion of political biology is Carl Schmitt's intuition regarding the inherently polemical nature of social and political constructs. In an oft-quoted passage from the *Concept of the Political* (1927), Schmitt claims:

> Words such as state, republic, society, class, as well as sovereignty, constitutional state, absolutism, dictatorship, economic planning, neutral or total state, and so on, are incomprehensible if one does not know *exactly who is affected, combated, refuted, or negated by such terms*. (1976, 30–31: my italics)

Is this insightful view also applicable to scientific concepts or expert claims (Nelkin, 1975; Turner, 2007)? Schmitt's answer would probably be that the decision to keep science or expert claims out of the political sphere, by rendering scientific statements non-polemical and proposing them as objective claims, is itself a profoundly political decision.

Although I am not arguing that notions of the gene, heredity, or human nature have to be seen on the same political (and therefore polemical) grounds as the concepts Schmitt analyzes, my political biology holds that in biology no major theory (e.g., heredity, human nature, nature versus nurture) was ever elaborated without implicit or explicit reference to political factors, and, once elaborated, every scientific position becomes a force affecting morality and politics, often in contradictory and ambivalent ways.

An *underdetermined* relationship

It is important from the very beginning to avoid a possible misunderstanding in my political biology project. I do not use the notions of political biology to embellish the crude claim that there is *a* one-to-one relationship between epistemology and political ideologies or any sort of logical necessary connection between the two. Quite the opposite. The relationship between epistemic statements and political values is full of counterintuitive and paradoxical developments. Theories usually associated with the right, such as hard hereditarianism, have played a more complex political role, and while Lamarckism is supposedly progressive, it harbors an often-overlooked potential for racism (see Stocking, 1968; Bowler, 1984).

Science and politics are in a relationship of reciprocal underdetermination. This means that any one scientific statement is logically consistent with multiple political outcomes. This is a well-known truth in the history of science, and that of human biology in particular. As Robert Proctor has written, "The history of science is often a history of confusion and ideologies often come in cumbersome packages" (2003: 226). Diane Paul, who has been so influential in showing eugenics' penetration of the left, agrees: "Scientific theories are socially plastic; they can be and frequently are turned to contradictory purposes" (Paul, 1995: 44–45).

So why bother at all with the idea of a political biology? If, in human heredity debates, nearly any political claim can be drawn from any scientific statement, and "many theories can respond adequately to the same social pressure" (Bowler, 1984: 260), why think in terms of political biology? I will discuss more deeply some of these objections in Chapter 4. For now I note that even if any scientific statement can have multifarious political applications, alternative scientific styles and doctrines actively color and frame different political positions. As we will later explore in greater detail, both Lamarckism and Mendelism were employed for right- and left-wing arguments, but to be a left-Mendelian implied a different doctrine and order of priorities than could be expected of a left-Lamarckian. H. J. Muller and Paul Kammerer, left-Mendelian and left-Lamarckian respectively, were both excited by eugenics, and both were materialists and believers in evolution. But they held different views, and the vocabulary of science definitely shaped and consolidated this pluralism. This is why left Mendelians and left Lamarckians, holding alternative views of heredity, clashed to the death in the 1920s in Soviet Union. And, going now on the right of the political spectrum, when early twentieth-century right-wing Mendelians and right-wing Lamarckians spoke of degeneration, they did not mean the same thing: better stocks being outbred by worse ones for the first, pathogenic environmental effects on human heredity for the latter.

Political biology tracks this ambiguous and contingent space where science is mobilized on behalf of politics and helps us understand the variable coloring of political options and worldviews via scientific vocabulary.

A second heuristic value of using the theoretical lens of political biology is genealogical – to explore periods of turbulence before scientific concepts become fixed to social worldviews (the "natural alliance" between, say, eugenics and conservatism). During these times of uncertainty, individual scientific facts are open to exploitation by many ideologies, and a single epistemic assumption covers "a rather wide range of political sentiments" (1981), as historian of science Loren Graham puts it. This open and undetermined potential eventually gives way to a "crystallization of values." (1977, 1981)

Expanding on Graham's dynamic notion, this book's effort is to move back diachronically to that foundational moment, which I saw occurring for debates on human heredity in the late 1920s before crystallization occurred. It is only after that moment that two apparently natural politico-epistemological alliances became fixed: the one between the left, the environment and nurture; and a second between (hard) hereditarianism, and the right. However this was only possible because, as I will argue, alternative solutions were destroyed and neglected, their promoters silenced rather than vanquished by others' superior theories. Although it is fair to recognize a pre-history of these associations in the nineteenth century (for instance in John Stuart Mill, see Paul and Day, 2008) nurturism and environmentalism *need not* be as such leftist values, nor biological heredity a rightist one. An emphasis on the primacy of the environment can always be, as I will show, one that highlights its pathogenic and destructive effects and bring to exclusionary discourses based on racial and social hierarchies. And a celebration of the power of hard heredity may be used as a political hope that "even after long-continued bad conditions, an enormous reserve of good genetic potentiality can still be ready to blossom into actuality as soon as improved conditions provide an opportunity" (Huxley, 1949). What we took for granted during most of the twentieth century was mostly the contingent effect of the way in which a certain controversy was closed, at least in Europe, in the interwar period.

In sum, political biology reopens closed connections between certain scientific statements and political values, connections that have been chopped off under the weight of historical stratification. This reopening seems particularly important right now, as a particularly neglected view of science – soft heredity – is being renewed.

Political biology as a triangle

Before moving to situating political biology in the context of other contemporary debates, a word on scientific language and its public uses. Political biology is not solely committed to a dialectics between epistemic and political facts in human heredity. There is a third component that is important to keep in mind, although it will be only marginally the focus of my book: the set of rhetorical resources employed by scientists as public figures. These are the metaphors and narratives by which scientific discourse becomes a legitimate public force (Gilbert and Mulkay, 1984). A substantial body of scholarship recognizes that "science is no exception to the rule that the persuasive effect of authority, of ethos, weighs heavily" (Gross, 1996; see also Knorr-Cetina, 1983; Gilbert and Mulkay, 1984; Lenoir, 1998; Ceccarelli, 2013). Language is a crucial heuristic and conceptual tool, and often the means by which the common-sense values of wider society are transmitted into scientific statements.

However, this appropriation, translocation, and re-contextualization of knowledge claims (Smocovitis, 1992) into the public sphere does not occur freely. Knowledge claims pass through an intense negotiation with the other two components of the political epistemological triangle: the constraints imposed by acceptable – that is to say, recognized – epistemic statements and available sociopolitical values. This triangulation with the epistemic and sociocultural context is often an unconscious one, occurring at the level of biases and presuppositions (Gould, 1981/1996), or, more properly, at the pre-intentional level of conceptual structures. Under certain historical conditions, given scientific practices or claims are deemed politically inopportune, embarrassing, or epistemically discredited. Scientists respond with "prudence" or perception of "danger" (Proctor, 2003), and some hypotheses – for instance, Lamarckism after Kammerer's suicide or at the peak of the Cold War – are marginalized as implausible.

The process whereby certain epistemic statements take center stage and others are silenced is often tacit and informal (Polanyi, 1958; Collins, 1974, 2010; Wynne, 1976; Bloor, 1976; Mulkay, 1979; Barnes, Bloor and Henry, 1996), not always assessable by objective criteria (Polanyi, 1958; Ravetz, 1979). As Jasanoff claims, science often proceeds by "incorporation of tacit cultural norms into the manufacture of credible evidence" (2012). This also means that what appears to be a rational process of knowledge growth is, if not a mere post-hoc rationalization, often and substantially driven by informal justification (Wynne, 1976).

To summarize, political epistemology, and in my case political biology, has to be seen as a triangle balancing the simultaneous tensions generated by the three poles of *epistemic statements*, *political strictures*, and *rhetorical and persuasive tools*. Each of these components is irreducible to the other two. Political epistemology is what occurs in the intermediate area between the tensions and negotiations of these three variables.

Political biology is not biopolitics

Biology, among all sciences, is a uniquely favorable terrain for thinking about this entanglement of politics and epistemology. Biology touches directly on the human condition: that the human being has been recognized since the beginning of Western thought as *zoon politikon* illustrates the profound intuitive intertwinement of our animal *and* political nature. As Foucault and Foucauldians have probably understood better than anyone else, biology is particularly prone to intense forms of politicization, albeit "beneath the official level of legal and constitutional discourses" (Marks, 2008: 98). This politics deals less with "legal subjects" and more with "living beings" whose vital activities become the target of biopower (Foucault, 1978, 2002, 2008; see Rose, 2007; Fassin, 2009a).

The supremely political nature of biological debates is most evident in the nature-nurture controversy (Pastore, 1949; Cowan, 1972a and b; Paul, 1998), as I discuss throughout this book. And the biographies of evolutionists and geneticists themselves are no less impregnated with political values. Nobel laureate H. J. Muller is a clear instance, with his claim to "have never been interested in genetics purely as an abstraction but always because of its fundamental relation to men" (quoted in Ludmerer, 1972: 37). But even when this relationship seems hidden, the interplay of political and scientific ambitions is nonetheless at work. As Dobzhansky wrote:

> Scientists living in ivory towers are now quaint relics of a bygone age. Nowadays, men of science must take note of outsiders peering at them and their work; more than ever before, their work and their writings are made use of, not only in economic, but also in social and political fields. Anthropology, the Science of Men, especially when concerned with the study of human races, is particularly vulnerable to misuse (1963).[4]

The quotation is particularly significant, as it is taken from Dobzhansky's polemical review of Carleton Coon's pro-racialist view of human phylogeny in *The Origin of Races* (1962). The epigraph from Simpson also speaks to this awareness on the part of evolutionary thinkers. Debates on heredity were explicitly and profoundly imbued with political metaphors. Galton made two overt political analogies when describing the relation between parents and offspring from the viewpoint of his new view of heredity: no longer a direct relationship as "between colonists and their parent nations" but a more "circuitous" one, as between "the representative government of the colony" and "the representative governments of the parent nations" (1876). In the middle of the 1950s, with molecular biology blooming, David Nanney opposed two views of the "cellular economy," the first of which equated the power of the gene, the "Master Molecule," to that of a "totalitarian government" (1957; see Keller, 1985; Sapp, 1987). It is pointing at this inextricable moment of science *cum* values that I will try to flesh out the notion of political biology in the book.

A little genealogy for political biology

There are several good models for framing and investigating the irreducible reciprocity of political and epistemic statements in the history of biology. Although Desmond and Moore (2009) do not use the label of political biology nor political epistemology, their recent work on Darwin is a good example of what I have in mind. Darwin's scientific position on the common descent of humankind, they argue, is strictly connected, both in its genealogy and in its effects, to his moral passion for anti-slavery. As the authors claim, Darwin's view of human ancestry has to be seen in strict relationship with his humanitarian beliefs. Darwin extends his unity-of-race thesis to a unity-of-life thesis, and his "abhorrence of slavery" is turned into a crucial step toward his evolutionary view. In the context of debates over race the "unity of descent" argument a plea for a universal consanguinity of the human species (Desmond and Moore, 2009).

Other examples of the profound interplay of epistemic and political positions and political arguments include Cowan's classic article (1977) on the genesis of the nature-nurture dichotomy in Galton's work; Shapin's (1979) study of competing interest in Edinburgh phrenology; Mackenzie's analysis (1982) of the link between statistics, hereditarian beliefs, and eugenics goals among biometricians; Robert Young's investigation of the interchange between Malthus's economy and Darwin's natural selection (1985); several critical works on the post-war "process of reciprocal legitimation" between Lorenz's nativism and Nazism (Kalikow, 1980; see also Griffiths, 2004); Jan Sapp's research on the struggle for authority in genetics (1987) and various political infiltration of embryology, cell theory and heredity debates (2003); Gilbert (1988) "Cellular politics" study on Just and Goldschmidt; Donna Haraway's reading of Washburn's paleoanthropology as a liberal-democratic prototype of the United Nations' universal man (1988); and Robert Proctor's analysis of the ideologically motivated post-war debates on human recency that delayed recognition of fossil hominid diversity (2003). Feminist studies of science have done more than any others in "waking up" (Martin, 1991) the sleeping language of politics, power, and gender in science and biology in particular (Keller, 1985, 1990, 2001; Haraway, 1988; Harding, 1986, 1991; Fausto-Sterling, 1985, 2000; Hubbard, 1990; Alcoff and Potter, 1993; Milam, 2010; Richardson, 2012; Alcoff and Potter, 1993). The critique of Eurocentric science in novel postcolonial approaches has been a logical development of this focus on the margins (Harding, 2006, 2013). Finally it is important to mention broader historical works on the ideological factors that produced sudden epistemic shifts in the heredity-environment debate (Cravens, 1978; Degler, 1991) and, in a parallel area, from racism to anti-racism (Barkan, 1992).

The trouble with biopolitics

It is only thanks to academic compartmentalization that this research on the politics of biology has only a tenuous connection to the broader debate that, especially after Foucault, proceeds amongst philosophers and social theorists under the umbrella of biopolitics. Although the term biopolitics is a "buzzword" with a complex history of usages, and many discordant sources (Lemke, 2011), it is undoubtedly Foucault's programmatic work that has represented a decisive spin to the current proliferation of biopolitical analyses. As is well known, Foucault interprets modernity as an immediate reflection of "biological existence... in political existence" (1978: 142):

> For millennia, man remained what he was for Aristotle: a living animal with the additional capacity for a political existence; modern man is an animal whose politics places his existence as a living being in question. (1978: 143)

In modern times politics becomes a source of dominion over the processes of life itself: power is biopower, "taking control of life and the biological processes of man-as-species" (2003: 247).

But while Foucault made a serious attempt to link historical analyses of biopower to broader overviews of, for instance, the emergence of biological rationality (see for instance his influence on François Jacob's

The Logics of Life), this link has gone almost completely lost in much political philosophy appropriation of biopolitics work since Foucault. After Foucault, and in a sense against Foucault, political philosophers of past three decades have cared little about historically inflected scientific renderings of life, which for Foucault had been the crucial target of modern politics.[5]

For instance, most of the influential political philosophers known today as biopolitical thinkers, from Giorgio Agamben to Antonio Negri, from Alain Badiou to Roberto Esposito have nothing relevant to say about the kind of history that is described here. Nor is biopolitics a useful entry point for any serious historiography of eugenics or racism. And even less is biopolitics, at this level of abstraction, illuminating of the possible conjoined genealogy of the social sciences as autonomous disciplines and the elaboration of notion of heredity between geneticists and eugenicists.

Most of the above-mentioned authors analyze biopolitics as a radically ahistorical, and universal notion with no time, no place, and no nuances. Bare life, state racism, biology and biologization, immunization are categories without a beginning or an end. This places us far from Foucault's injunction to avoid "deducing concrete phenomena from universals, ... and pass [instead] these universals through the grid of these practices" (Foucault, 2008: 3). This anti-Foucauldian style of inquiry of social theorists who today aim to expand on Foucault's analysis of biopower has been noticed by Rabinow and Rose:

> when Foucault introduced the term in the last of his *College de France* lectures of 1975–1976, *Society must be defended* (2002), he is precise about the historical phenomena which he is seeking to grasp. He enumerates them: issues of the birth rate, and the beginnings of policies to intervene upon it; issues of morbidity, not so much epidemics but the illnesses that are routinely prevalent in a particular population and sap its strength, requiring interventions in the name of public hygiene and new measures to coordinate medical care; the problems of old age and accidents to be addressed through insurantial mechanisms; the problem of the race and the impact upon it of geographic, climatic and environmental conditions, notably in the town. The concept of biopower is proposed after ten years of collective and individual research on the genealogy of power over life in the eighteenth and nineteenth centuries. (Rabinow and Rose, 2006: 199)

The lack of engagement with any serious analysis in the history of science and the current impact of scientific programs on the human condition produces fundamental flaws in many political philosophy view of biopolitics. To give just one example, though Italian philosopher Roberto Esposito (2008a; 2008b) is a bit more engaged in the history of science than are Agamben and Negri – and more fine-grained in his comments on the modern "flattening of the political into the purely biological" (2008a: 147) – he shows superficial knowledge of the history of twentieth-century biology and the dynamics of biologization. Thus he reads Nazi biopolitics as the paradigmatic case ("generalized to the entire world": 2008a, 2008b) that still articulates the essential relationship between individual, body, and state.[6]

Where biopolitical thinkers insist that the long shadow of Nazi biology lingers today, one of the historically documented arguments of my book is that, *as a structural reaction to Nazism*, post-war biology was tamed into a universalistic human rights–based framework. Even when biology in the mid-1970s drifted to more conservative and right-wing themes, for instance with sociobiology and later evolutionary psychology (see my intermezzo 2, after Chapter 5), what emerged was possibly an unpleasant phenomenon but still a fundamentally anti-totalitarian intellectual framework with human nature as a basic form of resistance to state intervention. This thesis may be questioned but not on the basis of an ahistorical view of biopower. Rather, the challenge for philosophers and social theorists would be to show in a detailed and historically situated way the epistemic and conceptual places where Nazi biology is supposed to persist. Other authors (Rose, 2007; Rabinow and Rose, 2006) have insisted in the change from a politics of death to one of life in contemporary biopolitics marked by the emergence of a molecularization of life processes, following a line of research different from the one at stake in this book. Again, biopolitical theorists may be unhappy with this, but to be credible, they need to dirty their hands with the concrete settings where biopolitics enters the mundane transactions of actual politics.

The project for a political biology is meant to offer an alternative to the frustration with the kind of philosophy and social theory which especially after Foucault (and alas, oblivious of Foucault) has indulged in abstractions over biopower. To make biopower more analytically tractable, the best antidote seems to me a profound commitment to approaches like historical epistemology and historical ontology. These may help bring the notions of biopolitics down from the heaven of abstraction to concrete historical practices and situated material contexts, where the form of bios has been made and remade over history, against the "firmament of ideas" (Foucault, 2008; see Meloni, 2010).

Why history matters to political biology: defining three epistemic eras

The relationship between biology and politics is not a night where all the cows are black. Far from it. In the post-Kuhnian world where we live, we know that "science is in time and is essentially historical" (Hacking, 1983: 6), and "epistemological concepts are not constants, free-standing ideas that are just there, timelessly" (Hacking, 2002: 8; see for post-positivism Zammito, 2004). Today's concepts and meanings are the ossified results of now pacified intellectual battles (Daston, 2000). We need history to see how the social and cultural structures that we take for granted are actually controversial, historically contingent, and therefore open to alternative possibilities.

Like the purloined letter in Edgar Allan Poe's short story, these structures, because of their proximity, are difficult to see. Historian of science Lorraine Daston calls this critical task an excavation of our most deeply held intuitions, a kind of "history of the self-evident" (2009). I draw on two methodological approaches to excavate the present: historical ontology and historical epistemology.

Building on Foucault's notion of a "historical ontology of ourselves" (1984), Ian Hacking (2002) developed a conceptual framework to describe how certain objects, in my case scientific statements, come into being. Hacking defines historical ontology as research concerning "the ways in which the possibilities for choice, and for being, arise in history." In the course of history, new spaces of possibility open and "create the potentials for 'individual experience.'" Daston's study of the effect of the camera on the notion of objectivity (1991a/b) is one example of the historical ontology or biographies of scientific objects (Daston, 2000; see also Daston and Park, 1998; Daston and Galison, 2010).

Historical epistemology (Lecourt, 1969; Daston, 1994; a critical review in Gingras, 2010) describes a similar effort to uncover the contingency and contextual dependence of scientific statements. In the wake of Foucault and Georges Canguilhem in particular, epistemology is turned away from a normative and ahistorical project about what counts as knowledge and toward a reflection on the historical conditions in which knowledge arises, the context "for objects to be made into objects of empirical knowledge," as Hans-Jörg Rheinberger (2010) puts it. Tim Lenoir (2010: XII) writes:

> In contrast to earlier traditions in the philosophy of science that treated truth as independent of the context of discovery and the history of scientific knowledge as a linear, progressive march in the elimination of error, asymptotically approaching nature, historical epistemology treats knowledge as historically contingent and focuses on uncovering the condition of possibility and fundamental concepts that organize the knowledge of different historical periods. (see also Rheinberger, 1997, 2005)

One message from this historical epistemological approach is that, contrary to the Wigghish illusion of a cumulative and linear continuity between past and present, scientific constructs are better investigated by using notions of dispersion, epistemic break, switch, and caesuras. Much of this is in principle known to (epistemologically aware) social scientists, but how do I apply in practice these ideas to my story?

Three biological eras

Following this nonlinear approach, I isolate three different political-epistemic regimes in the long twentieth century of biology, each involving specific articulations of the relationship between biology and politics and each separated by more or less visible caesuras.

Each of these eras is a "space of possibilities" (Hacking, 2002: 23), marked by specific styles of scientific reasoning and interpretative repertoires (Hacking, 1992; Harwood, 1993) and a cluster of shared cognitive and non-cognitive norms and values. Knowledge in each era, expanding on Fleck (1935/1979),[7] is not given by a dual relationship between the researcher's "individual consciousness" and its object, but decisively triangulated by the broader (collective) set of social values of the scientific community. Thinking in terms of styles invites us to see scientific thought, in each period or country, as "patterned (...) not simply a hodge-podge of unrelated attitudes" (Harwood, 1993). There is often a high level of idealism or psychologism in the definition of a style of thought (*Denkstil*), a scientific themata (Mannheim, 1953; Holton, 1978) or even around the much abused word "paradigm". I insist instead on the material scaffolding of truth via power, a co-production occurring mainly at a pre-individual level. In each era, political strictures filter and stabilize certain scientific statements (hardening them into accepted truths) and marginalize or silence others as epistemically possible but practically nonviable. A profound work of tacit negotiation occurs at this level. Inconvenient truths, such as radical eugenic measures or overt racism in post-WWII times, are more likely to reappear in the form of private exchanges between scientists or as "provocations" expelled or quarantined by the mainstream scientific community. But this politics is not just coercive. It also has productive, generative effects, eliciting a readiness for perception amongst scientific practitioners (Fleck, 1979) and making possible the reconstruction of scientific facts through certain scientific statements in line with the horizon of what can be said (for instance on individual freedom, race, gender and class) in a given epoch (Fleck's version of this would be that thought style "constrains the individual by determining 'what can be thought in no other way'" 1979).

Stabilization, marginalization, production and reinforcement of scientific statements do not occur in a mechanistic way. Rather, their effects follow a Gaussian curve of distribution; eras are not closed nor exclusive; what changes between them is the likelihood that certain scientific statements appear. I focus on this structural level, rather than on the level of individual scientific claims. As Foucault wrote in *The Order of Things*:

> At any given instant, the structure proper to individual experience finds a certain number of possible choices (and of excluded possibilities) in the systems of the society; inversely, at each of their points of choice the social structures encounter a certain number of possible individuals (and others who are not). (Foucault, 1970: 415)

In practice, in order to demarcate the different political-epistemic eras in my political history of human heredity, I use the following criteria: (1) whether heredity is conceptualized as hard or soft (i.e., whether both options are still available or one has entirely marginalized the other, that survives only as a "remnant" in Fleck's term); (2) how this conception affects the construction of the nature-nurture and biology-society border and, as a consequence, the relationship between the life and social sciences; (3) the intensity with which biological knowledge is applied to human affairs and its contribution to broader political values and narratives. Finally, I find useful to assess overall (4) how healthy Lamarckian speculations are in each era, a factor that usually is a good indicator of several of these variables (see Figure 1.1).

Each of the three eras so demarcated has its own scientific exemplars, its own politics and shared ethos, and uses forms of indoctrination and conformity to produce consensus amongst scientists. Whereas other scholars have emphasized the key function of research technology and experimental system to define historical settings in the history of heredity (Müller-Wille and Rheinberger, 2012), I rely on the political and value constraints that elicit, shape, and constrain the distribution of claims and

	1800 to Galton/Weismann *Prequel*	Galton/Weismann-1945 *The Story/1*	1945–1990s ca. *The Story/2*	2000– *Sequel (Postgenomics)*
Biology	Soft Heredity	Hard Heredity	Hard Heredity	Soft Heredity (Epigenetics)?? Extended heredity
Politics	Sparse Applications before 1859 Social Darwinism (after 1860s) Individualism, Optimism, Imperialism	Into the Wild (after 1900) Radicalism Social Engineering Utopia/dystopia: from the far-right to the far-left Degeneration anxieties	Democratized Four Pillars Universalism of Human Rights	Still Un-crystallized
Social Sciences	Evolutionary/Lamarckian Themes in Sociology and Anthropology (Spencer etc.)	Autonomization Via hard-heredity: Kroeber (via Weismann)	Frontal Fight or Inter- Disciplinarity (not challenging the divide though)	Entanglement Biosocial and Biocultural Investigations
Society / Biology Divide	Not existing The Social is Embedded in Biology (Habits Turned into Instincts)	Created Sociocultural is superorganic	Hardened 'Biology can't explain the Social (fallacy) 'Society can't explain Biology'	Collapsing? Biologizing the Social Position Socializing Molecular Biology
Citoyen Lamarck	Alive and Well (even Darwin...)	Wounded and Killed Golden Age ends (Gliboff) (Kammerer's suicide, 1926)	Nailed in the Coffin (Crick's Dogma Molecular Biology, 1958/1970)	Back from the Dead (Jablonka and Lamb, 2005)

Figure 1.1 A political epistemology of human heredity

their possible permutations in each era. This notion of shared ethos has resonances with that of moral economy, understood here as the "network of production, distribution, appropriation, and use of values and affects in society." Both aim to capture "the moral – including emotional – traits of the public construction of social problems, revealing or highlighting the values and affects driving ideological discourses, political decisions, and social mobilizations" (Fassin, 2015; see also 2009a). Finally, it is important methodologically to bear in mind that given my goal to contribute to a political biology (rather than a "political medicine" or "social policy"), my specific interest in delimiting these different thought-styles and eras is in how the professional community of biologists, or biologically-oriented intellectuals (i.e., authors who showed a theoretical interest toward debates on human heredity), rather than doctors, educators, novelists, or clinicians with no theoretical interests in heredity, contributed to biological knowledge and debates. This caveat about which "thought community" (*Denkgemeinschaft*, Fleck, 1979) selecting in the exploration of the politics of human heredity is particularly relevant when looking at eugenics and its periodization across the century. The methodological choice to analyze eugenics within "structures of disciplines and professions" (Pauly, 1993: 138) can help delimiting the overwhelming eugenics debate from the many other professional communities that animated the discussion and the practices around the world – doctors, educators, social reformers, social hygienists – in the global experience of eugenics as an international phenomenon. With this methodological detour in mind we can anticipate now the three politic-epistemic moments of my narrative.

The first era (1900–1945): from hard heredity to Nazi "barbarous utopianism"[8]

Although there is a prequel to this story with Galton's and Weismann's contribution to the creation of hard heredity in the late nineteenth century (Chapter 2), I take 1900 as a symbolic starting point for the epoch of intense, reckless politicization of biology. The first four decades of the twentieth century are characterized by the aggressive application of biological arguments to society: It is the time of eugenics and its plan to intervene politically and managerially upon the most intimate kernel of human nature, its germ-plasm, to take control of the genetic future of human populations via artificial selection (Chapter 3).

This is also the time when the social sciences, following a symmetric trajectory, would leave the biological in the hands of geneticists. Social science increasingly focused on nurture, the super-organic, leaving nature to biologists. The increasing discredit of biosocial Lamarckian explanation is part of this emancipation of the social, but also profound political reasons explain this break. Kroeber (intermezzo 1) is exemplar of this gesture in which (genetic) biological heredity and (disembodied) cultural heritage were radically dissociated.

At least until Kammerer's suicide in 1926, there were soft-hereditarian alternatives in eugenic thinking. The 1910s and 1920s decades enjoyed a rich variety of political-epistemic options before the crystallization of values occurred in the 1930s. Mapping these, included a significant strand of leftist hard hereditarians in Soviet Union, helps us to understand the polysemic and pluralistic character of eugenics not only as a movement spanning right to left but also from hard to soft heredity. This is why I speak of a Political Quadrant in eugenic debates (Chapter 4).

In conclusion, I claim that in the first era, 1900–1945, the varieties of eugenics were less united by a direct politicization of a particular theory of heredity than by a particular ethos. This ethos displayed four main features: the flattening of the notion of the human into its merely biological dimension, the view (often expressed in utopian language) that the future of human evolution had to be directly controlled by human efforts, a view of science as morally neutral and appropriate for inclusion in political and ethical debates, and the subordination of the good of the individual to the health of the species or race.

The second era (1945–2000): the democratization of biology and the hardening of hard heredity

A second historical period runs from 1945 to, approximately, the end of the twentieth century. The significance of 1945 as point of transition in the politics of biology remains controversial. Eugenic policies in country like the U.S. or Scandinavia remained well in place, and racism certainly did not disappear overnight. However, I will argue that after that date (with a process somehow started since the late 1930s), the politics of biology abandoned the bold social engineering spirit of interwar eugenics and repositioned itself successfully within a liberal-democratic framework. The radicalism of pre-1945 biology, right and left, was to be made compatible with the exigencies of the new liberal-democratic, post-totalitarian, universal human rights framework led by the American superpower, itself faced by an internal problem of racism and external problems of legitimacy in a rapidly decolonizing world. The modern synthesis of evolutionary biology, which was elaborated and consolidated between the late 1930s and culminated with Darwin's Origin of Species' centennial celebration in 1959, offered this moment of democratization a fitting universalistic and liberal-democratic framework. The work of Dobzhansky in particular epitomizes this novel center-leftist and liberal-democratic spirit. Obviously this repositioning was not immune from tensions and contradictions, as the debate around the UNESCO statement on race illustrates (Chapter 5).

However I disagree with arguments claiming that the persistence of race and eugenic themes implies that post1945 democratization of biology was just of façade. A politics of appearance and presentation to the public was part of the story but significant and substantial conceptual transitions came along with it. In Chapter 6, I focus on four of them that characterized the post-1945 scenario: (1) the classical-balance controversy (Dobzhansky versus Muller) and its implications for eugenics and human heredity; (2) the construction and rebranding of evolutionary thought in terms of population thinking and its visible disassociation from previous typological views of race; (3) the emergence of the idea of a human uniqueness and a specific cultural or psycho-social domain "exceeding" the biological dimension; and, (4) and the change in the political economy of nature from a group-oriented, holistic thought-style in biology to a fully individualistic view of biological processes. These four sites of transition were communicated to a wider public by major evolutionary thinkers as demarcating a good, fully democratic science from a potentially totalitarian one, which they made look now as a relic from the past. In parallel to this fourth transition, the hard/soft heredity controversy had a moment of (apparent) closure with Crick's central dogma of molecular biology (1958) that claimed that biological information could go only in one direction, from DNA to protein and never in reverse. Crick's dogma was taken as the "final nail in the coffin of the inheritance of acquired characteristics" (Mayr, in Wilkins, 2002). These important transitions helped stabilize the picture of post-1945 biology, radically decreasing the number of tensions (for instance typical of interwar biology) to which the politics of human heredity could be exposed. In practice there was no longer soft-heredity, nor the space to go too much on the right or the left of the political biological spectrum.

This new democratic framework proved tenacious even when theoretical assumptions of interwar eugenics returned much later in the form of sociobiology and evolutionary psychology. Faced with this rightward shift in the politics of biology, after the 1970s, the global democratic framework prevailed, turning sociobiology and evolutionary psychology into a tamed conservatism without eugenics. Sociobiology and evolutionary psychology's notion of human nature were anti-totalitarian and opposed to social engineering: Nothing could be more distant from a eugenic mentality.

The third era (2000–): welcome to postgenomics

In the last two decades or so, some of the pillars of the modern synthesis have started to crumble under the weight of a new postgenomic view of biology, of which the reactive genome and epigenetics are the most prominent examples. Francis Crick's dogma of molecular biology, which was conceived to put Lamarckism to rest, has come under increasing scrutiny.

Postgenomics does not just come after genomics; it is a different style of reasoning. It implies an unprecedented temporalization, spatialization, permeability to material surroundings, and plasticity of genomic functioning. There are many entry points to distinguish between genomics and postgenomics but I take the return of soft heredity, driven by the ascendancy of an epigenetic model of explanation, as the clearest marker for a postgenomic paradigm shift. The epistemic conditions that divided the social sciences from the materiality of biology have been significantly undermined. Until the 1990s, significant programs in the social sciences could play freely in the non-biological field of culture. Now, the new "social biology" that emerged with postgenomics has challenged this social-biological boundary. Politically speaking, the postgenomic scenario is *terra incognita*. It is full of potential but also presents significant problems, including the possibility of a new racialist and classist discourse in biosocial language. Epigenetics brings regeneration and degeneration, typical of interwar Lamarckian disputes, to the fore once again. Debates of a century ago, which seemed silent or relegated to the interest of the archivist and the intellectual historian, are again on center stage, albeit in the mutated epistemic forms of informatics-driven twenty-first-century life sciences. In the last two Chapters (7 and 8), the postgenomic era, I move in the insidious and rapid-changing terrain of our present, with its hype, speculation and scientific uncertainty, certainly a risky move. However this is not a good reason to shy away from some of the key questions that the present time is raising before us. For instance, how the politics of a new plastic biology will contribute to the understanding of social inequalities reproduction? What about epigenetic models of health and disease, discourses about the normal and the pathological at the time of epigenetic markers? How race, class and gender are being remade by the new postgenomic language of developmental and epigenetic differences? Insidious and perhaps premature (the skeptic would say) questions, but ones that any cartography or ontology of the contemporary has to try to answer, sooner or later. The alternative is to move blindly through our present as if we were before an inert mass of facts with no connection or historical direction.

2
Nineteenth Century: From Heredity to Hard Heredity

Against lines of descent

The construction of twentieth-century evolutionary thought, culminating in the synthesis of genetics and Darwinism, was not a cumulative and smooth process. It was, rather, an unsettled and complex contest among competing evolutionary frameworks. What we know as the most distinctive features of this twentieth-century synthesis – the full recognition of the role of natural selection, the emergence and consolidation of hard heredity, the "rediscovery" of Mendel, and the integration of genetics with natural selection – did not follow from "intrinsic logical necessity" (Sapp, 1983) but from epistemic ruptures and conceptual setbacks, political controversies and struggles for authority among diverse modes of knowledge (Sapp, 1987), and national styles of thought (Harwood, 1993).

The triumphant paradigm of twentieth-century biology is associated with the two iconic figures of Charles Darwin and Gregor Mendel – the former the father of evolution, the latter of the new view of heredity later identified with the gene. But this is mostly a teleological reconstruction. Others refined their ideas and made them compatible, giving rise to the modern evolutionary synthesis.

Take Darwin's role in the spreading of evolutionary views, for instance. In the second half of the nineteenth century, the effect of Darwin's work (evolution via natural selection) was that of a "catalyst" for a vast number of non-Darwinian views of evolution (Bowler, 1988). These evolutionary but non-Darwinian views looked at the adaptation or transformation of living forms without considering natural selection and often preserving a teleological view: idealism, progressionism, typology, orthogenesis, saltationism, recapitulationism, and various branches of Lamarckism, for instance. As Peter J. Bowler writes in *Theories of Human Evolution* (1986), "The *Origin of Species* precipitated the conversion of the scientific community to evolutionism, but did not necessarily dictate the structure of the theory to be accepted" (42). The persistence of non-Darwinian and developmental ideas in late nineteenth-century evolutionism, which Bowler calls the "Non-Darwinian Revolution" (1988), demonstrates that the decades following the publication of the *Origin* "did not exhibit a totally new world view, but more properly, an updated version of the old one" (Bowler, 1989: 49). And the endorsement of an evolutionary framework did not necessarily imply the recognition of Darwin's main tenet of natural selection, rather "must be dissociated from the process by which the selection theory came to dominate biology nearly a century later." "Recognition of this fact" Bowler continues, "forces us to reassess the impact of Darwin's ideas in his own time" (2009).

Darwin's role in making the modern view of heredity, which rejects Lamarckism and culminates with genetics, is also indirect. Darwin lacked the concepts that make possible hard heredity: none of its major dichotomies – nature and nurture, germ plasm and soma, genotype and phenotype – was to be found in Darwin's worldview (Johnston, 1995), although it is fair to say that he prefigured with his language some of them (Keller, 2010). Hard heredity, however, required more than mere prefiguration; it stood in need of a new epistemic space, and in that space, Darwin's notions of reproduction, variation, and heredity had to be, if not rejected, certainly profoundly revised and adjusted.

In *Variation of Animals and Plants under Domestication* (1868), Darwin famously offers the culmination of a quarter century's speculation on heredity (Olby, 1966). His theory, the provisional hypothesis of pangenesis, bears elements of what was then the cutting edge alongside ancient suspicions.

Darwin posted that every body cell "throw[s] off minute granules which are dispersed throughout the whole system." These "minute granules," or "gemmules,"

> when supplied with proper nutriment, multiply by self-division, and are ultimately ,developed into units like those from which they were originally derived... [Gemmules] are collected from all parts of the system to constitute the sexual elements, and their development in the next generation forms a new being; but they are likewise capable of transmission in a dormant state to future generations and may then be developed. Their development depends on their union with other partially developed or nascent cells which precede them in the regular course of growth. (Second edition, 1875)

There are undoubtedly anticipatory aspects of a modern view of heredity here. Like hard-hereditarian successors, Darwin envisioned heredity as the circulation of units of life. He favored the notion that "the true carriers of hereditary properties are not parents and their respective offspring, but submicroscopic entities" (Müller-Wille and Rheinberger, 2012: 38; Robinson, 1979; Gayon, 1998).

But pangenesis is also an old idea, traceable to ancient Greece (Johannsen, 1911; Zirkle, 1946: Blacher, 1982; Mayr, 1982). Many aspects of the theory look nothing like the hard heredity that the modern synthesis adopted. Abstraction of heredity from relations between parents and offspring, alone, does not make Darwin a hard hereditarian.

Where hard heredity demands strict separation between germ and soma, pangenesis relies on the opposite: direct communication between body cells and reproductive organs. According to Darwin, "Every part of the body threw off gemmules at various developmental stages." These would then be captured by the sex organs for eventual dispersal to the next generation. The hereditary process of transmission is therefore "contingent on the development of the organism" and there is no wall between transmission and "environmental influences." In the later terminology of hard heredity, gemmules are not "sequestered" from the environment (Winther, 2000).

The model of inheritance advanced by pangenesis holds that the germ cells consist of an "accumulated surplus" of the gemmules. This resembles the "overgrowth" idea proposed by nineteenth-century German biologist Ernst Haeckel, whereby parents achieved "material extension" to the offspring (Churchill, 1987). Because "the parent's body *manufactures* the particles from which the body of its offspring will be constructed," (Bowler, 1989: 25) pangenesis implies the transmission of acquired characters from parents to offspring. As Provine wrote: "The theory was designed so that the 'direct and indirect' influences of the 'conditions of life' might become embodied in hereditary constitution of the organism. If an organism were affected by the environment, the affected part would throw off changed gemmules which would be inherited, perhaps causing the offspring to vary in a similar fashion" (2001).

It is not surprising in this context that Darwin relied in many passages on a Lamarckian mechanism[1] through which habits become instincts and the effects of use and disuse are heritable. This can be seen as a further symptom of Darwin's resonance with a nineteenth-century developmental view of heredity (and a sign of difficulty in fully believing in the power of natural selection, as Thomas Hunt Morgan noted in 1903).

A final illustration of the gap between Darwin and hard heredity appears in an 1871 *Nature* exchange between Darwin and his younger half-cousin (via common grandfather Erasmus Darwin), disciple and sometime rival Francis Galton (see Gillham, 2001: 177 and ff.; Fancher, 2009; Bulmer, 2004a). Galton, soon to become a key figure of emerging hard heredity, sought at the time to put the validity of pangenesis and the existence of the gemmules to the test – an experiment that, at first, Darwin greeted. Between 1869 and 1871, excited by the possibility of applying his statistical skills to the new field of heredity, Galton transfused blood among different breeds of rabbits and examined their offspring to see if characteristics of the blood-donor rabbit were to be found in offspring of the blood-recipient parent, and therefore if "reproductive elements" circulated in the blood. The experiment was meant as "a direct and certain test" for the truth of pangenesis (Galton, 1871). He found no evidence, placing Darwin on the defensive. As the dispute between the two became heated, Galton (who had published somehow undiplomatically his negative experimental results in the *Proceedings of the Royal Society*) backed away from attacking Darwin's theory directly and formally, if unenthusiastically. "In the meantime," he wrote in *Nature*, "viva Pangenesis" (1871). However, this was the last hurrah for pangenesis by Galton. After that episode, which we can consider as a sort of parting of the ways, Galton's view of heredity developed autonomously and originally. His view of the stirp and his new theory of heredity, as we shall see soon, was just four years away. A demarcation from the famous half-cousin was obtained, with important consequences for the future.

It is fair to say, in sum, that Darwin's view of heredity did not make "a clean break" with the past (Bowler, 1989:63). He remained a "lifelong generation theorist" (Hodge, 1985), fascinated with growth and development in a manner typical of nineteenth-century biology (Winther, 2000).[2]

If Darwin was only indirectly the father of the new knowledge regime of hard heredity that came to dominate the twentieth century, so too was Mendel. This relationship is obscured by the stereotypical presentation of the man as an unnoticed and obscure genius, forgotten until his rediscovery in 1900. In fact he was known by his contemporaries within the context of debates on "speciation via hybridization," (Brannigan, 1979) rather than heredity and variability (see also Brannigan, 1981; Moore, 2001; Sapp, 2003). Moreover, Mendel's conceptual framework was scholastic and non-materialistic, based on Aristotelian notions like the "potential" (Kalmus, 1983; Callender, 1988; but see Hartl and Orel, 1994) rather than the intellectual structure and questions placed upon him after 1900, that is the discovery of material and particulate inheritance.

That is, Mendel was no Mendelian, and the long neglect of Mendel a "pseudo problem" (Olby, 1979; Bowler, 1989). Rather than being forgotten and rediscovered, this Augustinian monk who belonged to Linnean tradition of hybridists engaged in debates on the genesis of new species (Sapp, 2003), was *reconstructed* according to a set of conceptual problems that were urgent for early geneticists (Brannigan, 1979). "Mendel's laws," historian of evolutionary thought Peter Bowler wrote, "remained unrecognized precisely because no one in the period before 1900 was prepared to consider a study of heredity that was so obviously divorced from the problem of growth" (1989: 110; Sapp, 2003; see for a different reading Falk, 1995).

Only when, with the key contribution of Galton (stirp) and later August Weismann (germ plasm), the decisive transition to hard heredity was made, radically disjoining individual development from reproduction (since "what happens in growth cannot affect the transmission of germinal elements" Bowler, 1989: 83), did the rediscoverers of Mendel – the botanists Hugo de Vries, Carl Correns, and Erich von Tschermak – take from his work a solution to a problem distant from his immediate concern. What they found was particulate inheritance and the existence of discrete hereditary units (factors), which had little to do with Mendel's work in the hybridization tradition. This "factorial" view of heredity (Morgan, 1917), as it was known in the early twentieth century, became the perfect match for hard heredity because these segregated factors could be transmitted unchanged and recombined across generations without concern for the organism's development and growth (Bowler, 1989). Now that a new epistemic space was constituted, Mendel could pose as its father and hero. As Weismann claimed, in spite of recognizing differences and tensions, "Mendel's law is an affirmation of the foundation of the germ-plasm theory" (quoted in Churchill, 2015).[3]

The modern synthesis's paradigm of genetics and natural selection therefore did not arrive as on a continuum of progress from Darwin and Mendel. Here we have another case of founding-father fables, to quote Sapp (1990). Their questions and conceptual frameworks could only be retrofitted to this paradigm teleologically, a move generally frowned upon by historians of science. Teleological thinking can be the source of many mistakes and erroneous assumptions, especially when the understanding of a revolutionary development such as the creation of hard heredity is at stake (Müller-Wille and Rheinberger, 2012).

By eschewing an easy path from Darwin and Mendel and instead focusing in detail on how this novel regime of heredity was made, we will be in a better position to understand the social implications and cultural legacy of that shift. From a flexible and imprecise concept, heredity would become a "rigid" entity, "quantifiable, explorable and researchable" (Cowan, 1985). With this reification and atomization of heredity (Müller-Wille and Rheinberger, 2012), a truly modernist work of purification from all the unnecessary tinsels that the early and mid-nineteenth century had put upon it, a new understanding of the relationship between humans and their biological substance was born. Its meanings were many and significant, and the ways to make a politics out of it complex, subtle, and somehow counterintuitive.

Heredity before hard heredity

Writing a history of heredity is a complex task. The construction of the modern notion of heredity is part of a broader transition that brought to the emergence of biology as a science of life differentiated from the natural history of the classical period (Foucault, 1970; Jacob, 1973). In this discontinuous passage, which brought the notion of depth and organization at the center of the new biological view, discourses and metaphors about reproduction and heredity started to replace older notions like generation. In this previous framework, as François Jacob famously observed, "Living beings did not reproduce; they were engendered. Generation was always the result of a creation which, at some stage or other, required direct intervention by divine forces." Until the seventeenth century, generation was a "unique, isolated event" with "no roots in the past", similar to "the production of a work of art by man" (1973: 19–20). From this rather undecipherable for us conceptual framework to the coming to the fore of the biological notion of heredity, with its emphasis on the circulation of a common substance across the various organism, there is obviously quite a large jump.

"Until the mid-eighteenth century," Staffan Müller-Wille and Hans-Jörg Rheinberger write in one of the most comprehensive historical studies of the field, "the term 'heredity' was absent from theoretical writings on the generation and propagation of living beings." Indeed, it was not at all clear what sort of thinking should be marshaled to understand the inheritance of traits: "Inheritance was not a unified biological object in early modern thought, but was rather a subject distributed across different and often disparate domains" (2012: 43). Views of heredity were scattered among philosophical commentaries, encyclopedias, medical and moral treatises, and other sources.

But while a systematic concept of heredity was only recently formed in the life sciences, hereditarian views of disease were known before modern times. Historians such as Carlos López-Beltrán have shown that the notion of hereditary disease, or *morbi haereditarii*, was frequently used in medical writings since at least the Middle Ages. But these hereditary diseases were viewed as exceptions, not the basis of a general theory about pathology. And, as López-Beltrán notes, the word "hereditary" mainly appeared in an adjectival form (López-Beltrán, 2004, 2007), leaving the nominal form exclusively to convey its juridical meaning, mostly in the context of regulating blood relationships and inheritance of property (Müller-Wille and Rheinberger, 2012, see also Johannsen, 1911).[4]

It is since the early eighteenth century medical literature becomes replete, even anxious, with interest in heredity. Already during the revolutionary period in France, for instance, early theorists of what would become biology developed and discussed hereditarian themes that anticipate what would later be known as eugenics. These scholars were concerned with, among other things, the biological perfectibility of the new revolutionary man (Quinlan, 2010).

López-Beltrán in particular (2004, 2007) has analyzed this early nineteenth-century French medical context, where the notion of *hérédité* started to be nominalized and investigated as a phenomenon in itself. At some point in the early decades of the nineteenth century, he writes, "French medical men and physiologists adopted the noun '*hérédité*' as the carrier of a structured set of meanings that outlined and unified an emerging biological concept." Increasingly after 1830, and especially after Prosper Lucas's *Treatise of Natural Heredity* (1847–1850), the traditional medical formula "maladies *héréditaires*" was being transformed into "hérédité des maladies," "hérédité morbeuse," and "héredité pathologique." "Heredity," Beltran writes, "had become an unquestionable explanatory tool, capriciously adaptable to all evidential patterns, and, underpinned by a very thick network of general reasoning" (López-Beltrán, 2004: 61).

A similar process was underway in Britain. John Waller has shown (2001, 2002, 2003) the proliferation during the Victorian period of "discourses on hereditary transmission and 'prudent' reproduction" (2001: 458) in popular and medical literature. By the 1880s, "heredity" acquired the status of a dominant and "ubiquitous catchword" in medical sciences.

This hereditarian language suffused and enabled Galton's discourse decades later. But, as Waller himself recognizes (2003), the terminology itself didn't make a new epistemic framework, that is, hard heredity. Hereditarian terminology proliferated amidst the most disparate and contradictory views of heredity itself (see also Churchill, 1976). Nineteenth-century medical thought is full of hereditarian aetiologies for scrofula (Lomax, 1977), syphilis (Lomax, 1979), gout, consumption (tuberculosis), alcoholism, and insanity, but physicians did not attribute these maladies to nature rather than nurture, or to innate instead of acquired characteristics. "Before 1900," Charles Rosenberg writes, "there was no question in the medical profession but that acquired characteristics could become hereditary, that alcohol, drugs, sub-standard living conditions would debilitate parents and result in their producing weak, degenerate offspring" (Rosenberg, 1967). Early notions of biological heredity envisioned a blurred, confused mechanism "beginning with conception and extending through weaning" (Rosenberg, 1974). That breeders continued to believe in telegony speaks to this lax and broad view of what is passed from one generation to another.

Alongside medical literature, evolutionary writings continued to grapple with the inheritance of acquired features. Western thinkers well before Lamarck[5] (Zirkle, 1946; Blacher, 1982) sought this mechanism, which allowed no division of heredity from environment. No serious opposition between the two was possible when it was believed "ongoing social processes could still affect heredity" (Stocking, 1968: 244) and habits or racial memories (what today we would call sociocultural experiences) could turn into biological instincts. Darwin's grandfather, the Enlightenment polymath Erasmus Darwin, viewed heredity "as the result of a malleable admixture of nature and nurture causes" (Wilson, 2007). Although a terminology of nature and nurture is traceable at least to the Elizabethan age (Shakespeare, Mulcaster), with parallels in non-Western cultural contexts, a truly antagonistic opposition between the two terms (and therefore between heredity and environment) does not appear anywhere before Galton's writing on heredity in the 1860s (Keller, 2010). The folding of heredity into a substance separated from its environment, including the somatic cells, and subjected to a different economy than those cells produced a conceptual earthquake. It was called hard heredity.

The epistemic rupture of hard heredity

A key step in the construction of hard heredity was made possible by a reconfiguration of the architecture of the cell particularly after 1870 (Churchill, 1968, 1987, 2015; Müller-Wille and Rheinberger, 2012). As

Churchill writes, "the ten years from 1873 to 1882 were extraordinary in the annals of biology. With respect to the development of nuclear cytology, the decade possessed all the excitement and multiple revelations of the decade preceding the discovery of the structure of DNA some seventy years later" (2015). A new generation of physiologists began to focus exclusively on the nucleus as "the bearer of hereditary material" (Sapp, 1983) and a mechanic explanation of heredity became a factual possibility (Churchill, 2015). Technological advances allowed finally microscopic observations of cell structure, birthing the discipline of cytology as an alternative to the speculations of zoologists, comparative morphologists, and paleontologists (Gliboff, 2011). Heredity thus became inextricably linked with knowledge generated by cytological research; cell theory and evolutionary theory had to be reconciled (Sapp, 1983).

This disciplinary and methodological shift had a crucial effect for the hardening of heredity. Importantly, it translated into a significantly reduced emphasis on the power of environmental factors in explaining how variations and evolution occurred. Botanists such as Carl Nägeli (1884) were the first to see that "the germ developed in accordance with its own organization independently of and unperturbed by the dissolved (nonorganized) nutrients taken in by the mother" (Nägeli, 1884, quoted in Gliboff, 2011). Nägeli, who had a Hegelian philosophical background (somehow evident in his very speculative approach), was one of the key figures of this transition to a novel understanding of heredity (Churchill, 2015).

The word "heredity" was now inextricably associated with biology, its legal usage considered antiquated. The first biological uses of "heredity" in English are attributable to Herbert Spencer in 1863, with his idea of physiological units as the basis for heredity, and Galton in 1868. In America, William Keith Brooks's *Law of Heredity* followed in 1883. The word "inheritance," used by Darwin until his death, went into disfavor and was replaced by "heredity," now considered more technical and specific.

Cytology's fruits in hand, scientists generated a slew of novel ideas about the workings of heredity at the cellular level. Beyond the already mentioned Nägeli, other major contributions came from Moritz Nussbaum, Eduard Strasburger, Oscar Hertwig, Albert von Kölliker, Edouard Van Beneden, Theodor Boveri, Hugo de Vries, amongst others (Churchill, 2015). However, for the sake of simplicity we could say that it is mostly the combined influence of Galton (although he followed a different, statistical, line of argument, unrelated to the study of the cell), Weismann, and later, already in the early twentieth century, Wilhelm

Johannsen that built the modern consensus, with heredity finally and fully sequestered from its environmental and developmental influences. In claiming this I have no intention to overplay the role of these three authors to the detriment of others (playing again the founding-father fable), nor overlook their different disciplinary traditions and discrepancies from the later genetic view of heredity (to comply with which, many adjustments were necessary). What I mean is that with these thinkers, a break with the past was achieved, and we enter the dichotomous epistemic space that later came to dominate most twentieth-century reflections. In it, what Galton called "nature" is progressively equated with the innate, and eventually, after 1900, genetics, while "nurture", now freed from biological influences, will become available exclusively to a social scientific gaze. In the future economy of the life-sciences/social-sciences divide, it is no overstatement to claim that this boundary process was probably the major and more long-lasting conceptual event of the last fifteen decades. In an arc of time that goes from Galton's essay on heredity (1876) to Johannsen's genotype conception of heredity (1911), the completion of this novel view was obtained.

The three architects of hard heredity: Galton, Weismann, and Johannsen

Francis Galton

A scientific polymath with contributions in the most disparate fields, Francis Galton (1822–1911) is an extraordinarily interesting intellectual figure, who drew his motivations from a complex interplay of social, political, and scientific imperatives. He perfectly embodied the versatile Victorian investigator, exploring everything from meteorology to survey-based social research, from fingerprinting to free associations later made famous by psychoanalysis. He also had a profound influence on the scientific debates of his time (Pearson, 1930; Forrest, 1974 and Gillham, 2001; see also Brookes, 2004). No interloper or dilettante, Galton had an important role in the journal *Nature*, to which he contributed regularly after its founding in 1869. He received many awards in recognition of his scientific achievements (Cowan, 1985).

But as much as he fit his era, Galton also represents an important step in the making of modern science: the emergence and legitimation of a new class of scientific professionals whose prestige was marshaled to influence moral reasoning and public policy in the name of "practical benefits" for the larger society (Waller, 2001: 102). These professionals sought, among other goals, to instill a more meritocratic ethos in social institutions, restructuring them along rational, scientific lines. A key component of their rhetoric was, naturally, an attack on rank privilege in the name of talent (Soloway, 1990; Waller, 2001), something that was also part of the discourse of British eugenics.

More importantly, this new class aimed to embody the sort of epistemic authority usually reserved to the Church (Waller, 2001; see also Turner, 1974; MacKenzie, 1982), as critics of the new scientism from Alfred Russel Wallace to G.K. Chesterston would polemically note over the years. Galton was one of the first of this group to argue that religion's dogmatism and philosophical non-quantitative approaches had to give way to a new understanding of social facts based "on the solid foundation of quantified measurements and statistical methods" (Gorraiz et al., 2011).

One of the key strategies of the ascendant scientific professionals was to put new methods of statistical analysis at the center of their efforts to quantify *everything*. This was an obsession for Galton, who famously translated into numbers the efficacy of prayers on longevity and illness recovery, and calculated the average number of brushstrokes needed for a portrait (1883). Some years later, the biometrician movement epitomized this emerging quantitative spirit, and Galton, who helped to establish the journal *Biometrika* in 1901, was considered its founding patron (Porter, 1988; Gigerenzer et al., 1989).

But Galton was not all observation and empiricism. His theory of heredity was shaped by extrascientific motives and social attitudes (Cowan, 1972b; 1977; Mackenzie, 1982; Bowler, 1989; Gayon, 1998). Many critiques have highlighted the tensions between Galton's science and his ideology, his profound intellectual legacy alongside his weak scientific grounds and his lax methodological assumptions. His hard-hereditarian view has been deemed mostly the "result of socio-political rather than biological imperatives" (Cowan, 1977), the direct effect of an "ideological bias that runs through his studies of heredity" (Gayon, 1998: 105). "Rarely in the history of science," Ruth Schwartz Cowan writes, "has such an important generalization been made on the basis of so little concrete evidence, so badly put, and so naively conceived" (Cowan,1977). This ideological bias has frequently been seen in terms of a classism, but the racist motives behind Galton's work, part of it crafted during his African explorations (Stepan, 1982; Cowan, 1985; Yudell, 2014), should not be overlooked.

This entwinement of science and politics is evidenced in two hugely influential vocabularies that originate with Galton: nature/nurture and eugenics. The former sought to answer questions about the different influences acting on the human world; the latter sought to address the problem of how to apply and implement this knowledge to the benefit of society.

In Galton's *English Men of Science: Their Nature and Their Nurture* (1874), nature is defined as what "a man brings with himself into the world." Nurture, by contrast, covers "every influence from without that affects him after his birth," the vague "circumstances and conditions of life." At this juncture, Galton made an explicit epistemic caveat: the expression "nature and nurture" was to be understood as just "a convenient jingle of words, for it separates under two distinct heads the innumerable elements of which personality is composed." He maintained that "no theory" was implied by the recourse to this terminology, but as Keller notes (2010) a very loaded theory, one that will sneak easily into our language, is inherent in the tacit "presupposition of disjunction on which conjunction rests."

Not by chance, the discourse on nature and nurture as separate ontological realms solidified both in Galton's work and after. These became quantifiable and distinctive domains, endowed with different degrees of power. Each time the two entities competed "for supremacy," nature unmistakably proved the stronger. As Galton claimed:

> There is no escape from the conclusion that nature prevails enormously over nurture when the differences of nurture do not exceed what is commonly to be found among persons of the same rank of society and in the same country. (1883)

The nature of the seed always prevails on the nurture of the soil, to use a metaphor that would become popular in eugenic thinking. Nurtural dispositions, although visible (social class, for instance), remain superficial; they are "wholly insufficient to efface the deeper marks of individual character." (1874)

In parallel to this hugely successful conceptual landscape of nature and nurture, Galton also coined in 1883 the neologism "eugenics," from the Greek *eu*, "good," and *genos*, "birth."[6] The next chapter will deal more fully with eugenics, but note for now the neat division between overwhelming (inborn) nature and subordinate (acquired) nurture, and the politicization of that division by eugenics, "the science which deals with all influences that improve the *inborn* qualities of a race" (Galton, 1904, my italics). After all, if an inborn nature, i.e. heredity, was what really mattered in human life, and this nature was deaf to the solicitations of classic social reform, some other form of political intervention (eugenics) had to be invented to rescue human society. In Galton's worldview the two concepts are inseparably linked.

Indeed, the epistemic and the political are so deeply interconnected in Galton that it would be fair to say his biology is almost always a political biology. Galton's political-biological motives are evident not only when he focuses directly on eugenic aims, but also from the beginning of his thinking about heredity, which looks very much like a "biopolitical dispositive" (Müller-Wille and Rheinberger, 2012).

Interestingly, in *A Theory of Heredity*, Galton overtly mentions a "political analogy" to describe the structure of the hereditarian relationship between parents and children. Here Galton's model is not the relationship "between colonists and their parent nations," but the "more circuitous and feeble" connection between "the representative government of the colony with the representative governments of the parent nations" (1876).

In another passage of *A Theory of Heredity*, a second political metaphor appears:

> We know that the primary cells divide and subdivide, and we may justly compare each successive segmentation to the division of a political assemblage into parties, having, thenceforward different attributes. We may compare the stirp to a nation and those among its germs that achieve development, to the foremost men of that nation who succeed in becoming its representatives; lastly, we may compare the characteristics of the person whose bodily structure consists of the developed germs, to those of the house of representatives of the nation. These are not idle metaphors, but strict analogies (...) worthy of being pursued, as they give a much-needed clearness to views on heredity.

Galton's work on heredity and its influences

Galton's first publications on heredity came in the 1860s, long after he had been working in a variety of other fields and developing political leanings. "Hereditary Talent and Character," published in 1865 in the popular *Macmillan's Magazine*, was an important starting point. That article was later expanded into the book-length work *Hereditary Genius* (1869), where Galton investigated the genealogies of some of the most distinguished English families (judges, statesmen, academics, and artists) in order to carry out a sort of experiment to "test of the existence of hereditary ability." The result of the investigation should have been unequivocal, Galton thought, celebrating "a nature which, when left to itself, will, urged by an inherent stimulus, climb the path that leads to eminence" if in no way be impeded by "social hindrances" (1869).

The two main sources of inspiration behind Galton's turn to heredity in the 1860s were Darwin's *Origin of Species* (1859) and Adolphe Quetelet's *Letters on the Theory of Probabilities* (1846).

The *Origin* made a huge impression on Galton, pushing him to investigate a new set of issues including evolution, heredity, variation, and selection. His handling of the eugenic problem of improving mankind by selective breeding was definitely a result of reading *Origin* (Gillham, 2001). However, this relationship was not just one of passive learning. As we have seen, Galton tested Darwin's pangenesis, suggesting frustration with his half-cousin's view on heredity and his willingness to challenge Darwin on the novel terrain where Galton himself was about to become an authority. And Galton himself pushed Darwin to deal more overtly with the human problem (Paul and Moore, 2010). *The Origin,* as we know, stopped short of discussing human origins. However, twelve years later, *The Descent of Man* did it.

Meanwhile, Quetelet's (1796–1874) "political arithmetic" and "moral statistics" were crucial for the making of the social sciences in the second half of the nineteenth century. His *Letters on the Theory of Probabilities* was first available in England in 1849 and helped Galton recognize the regular distribution of human characteristics along a curve, a phenomenon he called "the very curious theoretical law of 'deviation from an average.'" (1869)

There was a tension between these sources: Darwin's natural selection principle, with his idea of individual variation, and Quetelet's notion of variation as mere perturbation from the mean. This remained a conceptual problem for the next decades in the biometrician movement (see Gayon, 1998).

If Darwin and Quetelet were Galton's scientific influences on heredity, who or what were his ideological influences? His ideological stance was rooted in a fundamentally anti-egalitarian political philosophy, explicitly affirmed in *Hereditary Genius*:

> I have no patience with the hypothesis occasionally expressed, and often implied, especially in tales written to teach children to be good, that babies are born pretty much alike, arid that the sole agencies in creating differences between boy and boy, and man and man, are steady application and moral effort. It is in the most unqualified manner that I object to pretensions of natural equality.

In taking this position, Galton swam against the tide of Victorian culture. He dispensed with both the moderate environmentalism of mid-to-late nineteenth-century British social thought and the Lamarckian inheritance of acquired characters in a period when that doctrine was widespread. "I am aware that my views…are in contradiction to general opinion," Galton wrote (see Gökyiğit, 1994). It took three decades after *Hereditary Genius* for sociopolitical values to shift from the liberalism and environmentalism of the 1860s to a more elitist and pessimistic view of social change, not coincidentally making Galton's hard heredity more palatable to his contemporaries. Only at that point did his eugenic proposal become a conceivable reform idea.

Galton had no formal training in biology and this probably spared him both the preconceptions of biology and its confusions at a time where there "was no common agreement amongst biologists about the definitions of phenomena to be studied" (Cowan, 1977). In sum, both Galton's outsider status in British social thought and his lack of training were likely instrumental in distancing him from Lamarckian assumptions and from the ameliorant social philosophy these assumptions usually implied. He wrote too early about heredity to enjoy the cytological developments described above, but at least avoided being influenced by that admixture of nature and nurture that was the notion of medical heredity at the time.

Galton's most important contribution to the new hereditarian discourse lies in circumscribing a novel and autonomous space for heredity, newly conceived as a "physiological connection between generations…open to research and quantification" (Soloway, 1990: 23). His statistical approach was particularly important because for quite possibly the first time in the nineteenth century, heredity and variation ceased to be seen as alternative, antagonistic forces. The solution to this apparent contradiction was to look at the level of the whole population, where heredity and variability became the expression of a common mechanism (Olby, 1966; Gayon, 1998). Galton, it has been said, "was the first to clearly present the relationship between 'descent', 'heredity', 'variability' and 'reversion'" (Gayon, 1998) It is also true, however, that Galton only partly understood the population view of heredity and variation. He elaborated a law of ancestral type, which held that all individuals in a species inherit traits from distant ancestors who serve as the true expression of the species. The tendency of the species is to regress to this type, in spite of selection pressures. This implies that the balance of a population could only be altered temporarily. Galton therefore was committed to a sort of "racial heredity or inertia" that is "clearly opposed to the Darwinian theory of natural selection." He "ended up encouraging archaic and confused ideas that gave credit to the concept of racial constancy," Gayon concludes (1998).

This notion of racial constancy was embodied in the *stirp* (from the Latin *stirpes*, "root"). Galton divided hereditary space into two domains, one made of "inborn or congenital peculiarities that were also congenital in one or more ancestors," or the stirp, and a second of characteristics "that were not congenital in the ancestors, but were acquired for the first time by one or more of them during their lifetime, owing to some change in the conditions of their life. The first of these two groups is of predominant importance" (1876). The *stirp* denotes for Galton the

> total of the germs, gemmules, or whatever they may be called, which are to be found, according to every theory of organic units, in the newly fertilized ovum – that is, in the earliest pre-embryonic stage – from which time it receives nothing further from its parents, not even from its mother, than mere nutriment. (1876: 330)

With the notion of the stirp, Galton accomplished an important theoretical novelty. Darwin and other generation theorists coupled reproduction and heredity in a unitary process. Heredity was the effect of outgrowth from parents to offspring (Churchill, 1987, 2015). This direct transmission explained why parents and offspring resembled one another. But Galton saw it differently. As Kyle Stanford puts it:

> Parents and offspring might share salient characteristics not because the parents' tissues or other physical features themselves contribute materially or even causally to the formation of those of the offspring but instead because *both* sets of tissues, organs and features (with their shared peculiarities) are produced by *shared germinal materials*, of which identical or systematically related versions are invariably passed from parents to offspring. That is, the tissues of the offspring (produced by whatever intervening mechanism) might recapitulate salient features of the parent's not because the latter serve as causes of the former, but because they share a *common cause* in the hereditary materials found in a shared germ line ultimately producing them both. (Stanford, 2006; italics original)

Compared to generation theorists, Galton saw a "more circuitous and feeble" relationship between the hereditary material of offspring and parents:

> The stirp of the child may be considered to have descended directly from a part of the stirps of each of its parents, but then the personal

> structure of the child is no more than an imperfect representation of his own stirp, and the personal structure of each of the parents is no more than an imperfect representation of each of their own stirps. (1876)

This meant that parents' bodies themselves did not cause changes in offspring's characteristics but functioned as a vehicle through which, according to Peter Bowler, a "package transmitted intact from a whole series of ancestral generations" was passed on to the next (Bowler, 1989: 68).

As Galton wrote in 1865:

> We shall therefore take an approximately correct view of the origin of our life, if we consider our own embryos to have sprung immediately from those embryos whence our parents were developed, and these from the embryos of their parents, and so on forever.

Thus Galton established a concept of heredity as "a process independent of the productive capacities of the parent organisms." Heredity began to be thought of as "a continuous line of descent from generation to generation," not something manufactured by parents and transmitted to offspring. Through this autonomization of heredity, Galton "paved the way for the complete destruction of the developmental world view by challenging the link between generation and evolution that even Darwin had left intact" (Bowler, 1989; see also Müller-Wille and Rheinberger, 2012).

With Galton we have moved significantly away from the developmental worldview in which Darwin himself was immersed. However, the epistemic break toward hard heredity could not be completed merely by Galton's statistical and ideological assumptions. A second major contribution came from the enormous advance in biological research, especially in the study of the cell, that characterized Germany in particular between 1870s and 1880s. August Weismann was the most achieved champion of this novel wave of studies.

August Weismann

Celebrated by Mayr as "the greatest evolutionist after Darwin," (1985) August Weismann (1834–1914) won his influence by articulating the broadest and most structured view of heredity of the entire nineteenth century. This was based on the notion of a sequestered, continuous and immortal substance, the germ plasm. It was a key part of this new picture a systematic attack on Lamarckism, the inheritance of acquired characters and its related developmental view of biology, philosophies still very widespread in German evolutionism and biology, from Haeckel to Virchow. Trained as a physician, Weismann became an embryologist and a professor of zoology and later director of the zoological institute at the University of Freiburg in southern Germany, where he spent most of his life and career (he was born in Frankfurt on the Main, see Churchill, 2015). In the last decades of the nineteenth century, Weismann's name came to be identified with a transformation from the "original, flexible Darwinism" that could still make room for a "Lamarckian component in addition to natural selection" to a more "dogmatic" one (Bowler, 1988: 75;) in which natural selection was the exclusive and omnipotent source of individual variation.

A vivid and very popular part of this story was the famous experiment carried out by Weismann in 1887–1888, when he tested Lamarckian inheritance by amputating the tails of more than twenty successive generations of mice. Their offspring all had intact tails. These experiments were intended to disprove the theory that acquired mutilations could be inherited across generations (for instance, it was popular story that the shorten tail of the Manx cats of the Isle of Man was an outcome of such induced mutation). In truth, Weismann's experiment had more a symbolic than a scientific meaning. As Churchill, Weismann's main interpreter and biographer, recognizes "there could be much to criticize about this procedure, but Weismann argued not that he had disproved the inheritance of the induced bobtailed condition in mice but that it was impossible to prove such inheritance could *not* take place". What Weismann was doing was to push Lamarckians to offer experimental evidence, and "place the burden of the proof" on their side (2015). A move that was in the end very successful. Lamarckian inheritance, according to Mayr (1988), "never regained full credibility after Weismann's attack" (see for criticisms to this view, Jablonka and Lamb, 1995, 2005/2014).

It is clear from this example that, as with any iconic figure, Weismann was a symbol for something and at the same time a more complex character than the ideas that would be named for him. As we will see, Weismann and Weismannism have been given different receptions in different times and places, and the quality of his reception provides a good measure of the intellectual climate during different phases of the history of biology.

James Griesemer and William Wimsatt (1989) have for instance studied the conceptual changes in the understanding of Weismannism through an examination of the different visualizations of Weismann's work – the changes in the diagrams used to illustrate Weismann's doctrine of the continuity of the germ plasm (on which the impossibility of the inheritance of acquired characters was based) are very telling of the different conceptual contexts where Weismannism has been utilized. Weismannism achieved "peak simplification" during the 1960s and 1970s, when twentieth-century gene-centrism was on the rise (in authors like Richard Dawkins for instance), and looked to Weismann as a forerunner of this later view.

The difficulty in distinguishing between Weismann and Weismannism, however, also lies in the fact that from the very beginning Weismann understood himself and was understood as the proposer of a crucial turn in the re-conceptualization of heredity and evolution. Weismann was a polarizing figure that could be embraced or fought against, but could not leave things as they were before. Before Weismann, natural selection and Lamarckian inheritance were seen as concomitant factors in the process of selection differing only by degree, not kind (Romanes, 1899). Heredity was a pluralistic mechanism. After Weismann, the polarization between these two mechanisms – natural selection and the inheritance of acquired characters – became extreme, giving rise to a series of ideological fights. The term neo-Lamarckians and neo-Darwinians were both created after Weismann's first important works on heredity, between 1885 and 1888 (or 1883 and 1885) as an effect of this polarization.[7]

The heated debate with Spencer in the *Contemporary Review* in the early 1890s (e.g., Spencer, 1893a and b; Weismann, 1893b) is very representative of this clash between what, after Weismann, emerged as two irreconcilable worldviews.

Weismann's work

Weismann was trained in an epigenetic view and initially "rejected the idea that individual form emerges through the unfolding, or evolution, of pre-existent form in the inherited germ" (Maienschein, 2005). But by the time of his 1883 lecture "Über die Vererbung," he was turning to preformationism, which, contra epigenesis, affirms that mature individuals are just an unfolding of embryonic forms of themselves. And with the publication of *Das Keimplasma* in 1892 – translated into English in 1893 as *The Germ-Plasm* – the transition from epigenesis was complete:

> I have myself more than once abandoned a line of research undertaken in connection with the problem of heredity ... what I sought was a substance from which the whole organism might arise by *epigenesis*,

> and not by *evolution*. After repeated attempts, in which I more than once imagined myself successful, but all of which broke down when further tested by facts, I finally became convinced that an epigenetic development is an *impossibility*. Moreover, I found an actual *proof of the reality of evolution*.... It is so simple and obvious that I can scarcely understand how it was possible that it should have escaped my notice so long. (Weismann, 1893b, pp. xi, xiii–xiv; italics original)

(When Weismann speaks of tests, proof, and facts, he means it. Although many considered him only a theoretician, he emphasized what he called "reasoning supported by observation" and pioneered, experimental approaches to studies of heredity. Still, he believed, "Other than experimental methods may lead us to fundamental views, and an experiment may not always be the safest guide, although it might first appear perfectly conclusive" 1893: xiv, 137).

Weismann's idea of heredity was known as the theory of the "continuity of germ plasm" and was based on the assumed "existence of a special organised and living *hereditary substance*, which in all multicellular organisms, unlike the substance composing the perishable body of the individual, is transmitted from generation to generation" (1893: xi). The doctrine of the continuity of germ plasm is a fundamentally dualist one, based on a "contrast between the *somatic* and the *reproductive* cells" (1893: 183). The two cells, on this view, are subject to different economies. As Mary Jane West-Eberhard in a now-classic work explains, "The cells of the soma participate in growth and differentiation, but then they die, while the germline cells, set aside early in development, serve as an uncontaminated bridge to the next generation" (1993: 331).

Germline cells are sequestered from somatic cells, "beyond the reach of any variation that might occur in individuals of the species" (Jacob, 1973: 217). Therefore, in higher animals, no passage whatsoever between the somatic level (what is acquired during a lifetime) and the hereditary substance is possible (importantly the argument does not apply to unicellular organisms or plants). What logically follows is that "characters acquired by the adult body cannot be reflected in the germ plasm and thus cannot be inherited" (Bowler, 2009: 255). After Weismann, this sequestration became known as Weismann's barrier. Weismann himself conceived of this barrier with a linguistic metaphor: to suppose communication between the two systems "is very like supposing that an English telegram to China is there received in the Chinese language" (1904).

The continuity of germ plasm brought new precision and heft to some of Galton's intuitions about particulate inheritance, the autonomy of the hereditary substance, and the denial of the inheritance of acquired characteristics.

First, with respect to particulate inheritance, Weismann understood the germ plasm as a hierarchical composition of vital units or *ids*, "capable of growth and multiplication by division" (1893: 63). The chromosomes (a term coined by German biologist von Waldeyer-Hartz in 1888) could be considered "series or aggregations of ids" (1893: 67). Thus with Weismann "heredity started to be understood as the transmission of nuclear substance with a specific molecular constitution" (Weismann, 1891). He also reasoned *a contrario* that, if inheritance were not particulate but a process of blending, it would follow very different laws of recombination (Mayr, 1991). In defending so clearly particulate inheritance, Weismann's ideas supplied the necessary basis for the development of genetics.

Second, with regard to the autonomy of the hereditary substance, Weismann supposed "that no part of the parent organism generates any of the formative material which is to constitute the new organism." The hereditary material stood and lived by itself. So wrote the neo-Lamarckian George Romanes, one of Weismann's first critics. In Weismann's view:

> this material stands to all the rest of the body in much the same relation as a *parasite* to its host, showing a life *independent* of the body, save in so far as the body supplies to it appropriate lodgment and nutrition; that in each generation a small portion of this substance is told off to develop a new body to lodge and nourish the ever-growing and never-dying germ-plasm – this new body, therefore, resembling its so-called parent body simply because it has been developed from one and the same mass of formative material. (1899: 26, my italics)

Finally, Weismann turned to three pieces of evidence to refute Lamarckian inheritance and the notion that the environment can directly guide heredity. As Mayr describes, first, "there is no cytological mechanism that could effect such a transfer [of acquired characteristics] from soma to germ plasm." Second, there "are many adaptations that could not have been acquired by such an inheritance (for example, the soldiers of ants and termites)." Third, "all reputed cases of inheritance of acquired characters can be explained by selection" (Mayr, 1988).

As an explanation of evolution, the inheritance of acquired characteristics could be jettisoned because Weismann found a better source of genetic variability. The "sexual mode of reproduction" and chromosomal recombination supplied all that was needed, Mayr writes:

> It was the process now called "crossing over." If such a rebuilding of chromosomes during gamete formation (meiosis) did not exist, genetic variation (except for occasional new mutations) would be limited to a reassortment of the parental chromosomes. By contrast, chromosomal recombination has the consequence that "no individual of the second generation can be identical with any other... [in every generation] combinations will appear which have never existed before and which can never exist again." No one before Weismann had understood the extraordinary power of sexual recombination to generate genetic variability. (Mayr, 1982: 537)

An evaluation of Weismann

Underneath the image of Weismann as anticipator of twentieth-century genetics and father of hard heredity, a more nuanced historiographical tradition has established that, as with Galton, Weismann pioneered elements of a radically new vision of heredity while adhering to old developmental, embryological, and recapitulationist views that persisted until his last publications (Churchill, 1999; Bowler, 1989; Winther, 2001; see also Mayr, 1991). Not only there were tensions and discrepancies between Weismann's model of germ plasm and the genetic revolution that took stage a few years after Weismann's major publications, a bandwagon on which Weismann did not jump up (Churchill, 2015), but scholars have rejected also Weismann's supposed commitment to an impermeable germ plasm, claiming he "by no means implied that the nuclear plasm... was unable to acquire new qualities" (Müller-Wille and Rheinberger, 2012; see also Jablonka and Lamb, 1995; Logan and Johnston, 2007). According to J. Novak (2008) for instance, in 1895 Weismann "proposed a supplementary concept of 'germinal selection' that re-introduced the inheritance of acquired features." Also accounting for Weismann's later ideas, R.G. Winther (2001:164) argues

> Weismann's defense of external and developmental sources of germ-plasm variation blurred the strong distinction that biologists, particularly geneticists, were forging. Furthermore, such a source of variation seemed to many to be a form of Lamarckism that disturbed the morphologically and variationally sequestered sanctity of the causally-powerful germ-plasm. Hence, it appears that they reinterpreted Weismann in a manner suitable to their purposes. Such interpretative

> moves would also favor the advocacy of eugenics and hereditarianism by many biologists.

There is no doubt that, as Winther suggests, broader political pressures hardened the Weismannian dichotomy between non-heritable somatic variations and germ-plasm heredity into a broader ideology, which we can call, beyond Weismann himself, Weismannism. This ideology was at the heart of swelling eugenic and hard-hereditarian schools of thought in the early twentieth century. In Germany, Weismannism were used to justify the militaristic ideology of the ruling elites (Crook, 1994, see also 2007). Kelly (1981) equates the influence of Weismannism on German Darwinism with a shift toward racism and classism. Weismann's views, he claims "came very opportunely for those who were increasingly anxious about the security of their own class or race".

The most immediate and visible effect of his work was to strengthen the idea of the "all-sufficiency of natural selection": only selection acting directly upon the germ plasm can cause evolution. This view certainly was important in the emergence of the racial hygiene movement in Germany, just a few years after Weismann's main writings on heredity. Sheila Faith Weiss quotes one German "race hygienist":

> It was Weismann's teaching regarding the separation of the germ plasm from the soma, the hereditary stuff from the body of the individual, that first allowed us to recognize the importance of Darwin's principle of selection. Only then did we comprehend that it is impossible to improve our progeny's condition by means of physical and mental training. Apart from the direct manipulation of the nucleus, only selection can preserve and improve the race.

"For those who accepted Weismann's views with respect to both heredity and the 'all-supremacy' of selection," Weiss concludes, "eugenics was the only practical strategy to ensure racial progress and avert racial decline" (Weiss, 2010).

German eugenicists such as Wilhelm Schallmayer immediately understood the political implications of Weismannism. To them, as Loren Graham writes (1977):

> Weismann's view of autonomous germ-plasm graphically illustrated that what mattered from the standpoint of modern biology, was that individuals secure the preservation of the species through proper transmission of their genetic heritage (*Erbgut*), not that man make

> more comfortable either his own temporal existence or that of his fellow man. In fact, more comforts and better social conditions might actually be detrimental to the future of the human race.

Nevertheless, beneath the more ideological uses, if not caricatures, of Weismann's thought (Weismannism), his profound and long-lasting impact as an original thinker is at the core of many arguments of this book. Also the political uses of Weismann were very different, as it will be soon evident, from some of the nationalistic and bleak representations of a fatalism of the germ plasm in militaristic sauce. The Weismann's barrier could be used for progressive goals in authors as Alfred Russel Wallace, Alfred Kroeber, Yuri Filipchenko, and Julian Huxley.

From a scientific point of view, Weismann garnered reactions on every front. In France, England, and the United States, researchers established neo-Lamarckian institutions and a neo-Lamarckian experimentalist tradition that tried to counter his arguments (Rainger, 1991; Cook, 1999). Philosophers and sociologists, too, bristled. By insulating the germ plasm from the somatic cells, Weismann broke the popular alignment between hereditarian and sociocultural dynamics. The intellectual impact was traumatic for those who defended the possibility of inheriting acquired characters, which was seen as a source of moral progress.

Herbert Spencer and American neo-Lamarckians, in particular, fought back. In a series of articles in 1893 in *The Contemporary Review*, Spencer (1893a, b) questioned the adequacy of Weismann's panselectionism, the all-sufficiency of natural selection in explaining evolution (Weismann, 1893b). There were, of course, technical issues about how evolution could work if acquired characteristics could not be inherited. But more profound philosophical problems were at stake. As Spencer recognized, "A right answer to the question whether acquired characters are or are not inherited, underlies right beliefs not only in Biology and Psychology, but also in Education, Ethics, and Politics" (1893a: 488). As Romanes put it,

> The main ethical and political concern, was that no matter how many generations of eagles, for instance, may have used their wings for purposes of flight; and no matter how great an increase of muscularity, of endurance, and of skill, may thus have been secured to each generation of eagles as the result of individual exercise; all these advantages are entirely lost to progeny. (1899: 20)

Lester Frank Ward (1841–1913), a prominent American neo-Lamarckian social scientist and first president of the American Sociological Association, was similarly perturbed by Weismann, the new "great prophet of science." Like Romanes, he surmised that if hard heredity were true, social progress would be lost. How could it be otherwise if each generation's political, moral, and educational efforts were erased with the rise of the next?

> If nothing that the individual gains by the most heroic or the most assiduous effort can by any possibility be handed on to posterity, the incentive to effort is in great part removed. If all the labor bestowed upon the youth of the race to secure a perfect physical and intellectual development dies with the individual to whom it is imparted why this labor? If, as Mr. Galton puts it, nurture is nothing and nature is everything, why not abandon nurture and leave the race wholly to nature? In fact the whole burden of the Neo-Darwinian song is: Cease to educate, it is mere temporizing with the deeper and unchangeable forces of nature. And we are thrown back upon the theories of Rousseau, who would abandon the race entirely to the feral influences of nature. (1891: 65)

Another American neo-Lamarckian, the paleontologist Henry Fairfield Osborn, expressed much the same concern about the impact Weismann would have upon "the conduct of life."

> If the Weismann idea triumphs, it will be in a sense a triumph of fatalism; for, according to it, while we may indefinitely improve the forces of our education and surroundings, and this civilizing nurture will improve the individuals of each generation, its actual effects will be not cumulative as regards the race itself, but only as regards the environment of the race; each new generation must start *de novo*, receiving no increment of the moral and intellectual advance made during the lifetime of its predecessors. It would follow that one deep, almost instinctive motive for a higher life would be removed if the race were only superficially benefited by its nurture, and the only possible channel of actual improvement were in the selection of the fittest chains of the race plasma. (quoted in Rainger, 1991)

Thus, not only in politics but also in social science quarters, hard heredity was at first understood as merely a reactionary and exclusionary doctrine, a heavy obstacle in the route of moral amelioration. From the perspective of the social sciences of the time, strictly connected to an idea of moral progress, it offered few if any advantages.

However, Weismannism inspired less-intuitive moral corollaries, as well. While neo-Lamarckians were concerned about the futility of education and moral efforts, in a later period leftist Mendelians, suspicious of what they perceived as the dangers of Lamarckism, argued that the negative effects of the environment would be contained and even neutralized by an impervious hereditary substance (Alfred Russel Wallace was the first to see this point already in 1892). After all, if the good of education could not be attached to heredity, then the ills of unequal social structures also would be kept at bay. The debate would have enormous political and epistemic consequences. Although rarely recognized, Weismann himself saw in passing the complex political implications of his germ-plasm doctrine: "the hypothesis of the continuity of the germ-plasm" he wrote "gives *an identical starting-point* to each successive generation" (1891: 168, my italics). Weismann was probably more concerned with the biological potential that hard heredity leaves intact for each generation rather than with a full-fledged politics. But the broader political implications of this sentence had reverberations in the arc of the next century (when the philosopher Fukuyama made the point about the democratic value of the genetic lottery, he was actually rehearsing this Weismannian *topos*). There was a modernist message in hard heredity about autonomy of the individual and the breaking of the chain of the past, although the rhetoric of hereditarianism as political fatalism has often submerged this progressive view. As Benjamin Kidd, the author of *Social Evolution* (1894), commented with optimism after meeting Weismann in 1890:

> every new generation comes into the world pure and uncontaminated, so far, by the surroundings and life history of parents. (quoted in Crook, 1994)

This is one of the first recognitions of the democratic potential of hard heredity, on which we shall come back in Chapter 4.

Summary: What Galton and Weismann accomplished

Before proceeding, it may be helpful to offer a synthetic overview of Galton and Weismann's joint contribution. Together, they helped to construct the modern view of biological heredity, now subsumed into a universal mechanism (Bonduriansky, 2012). In this mechanism, heredity is internalized in a substance passed across generations, independent of both parents' contributions, environmental influences, and in general the "vagaries of life". Heredity, thanks to Galton and Weismann, became something inside us and beyond us, something that can now be calculated to determine its relative influence on individuals, as compared to rigidly distinct environmental factors.

In spite of their different points of departure – Weismann an embryologist and a zoologist; Galton a polymath possessing statistical acumen – there is a significant correspondence between the two authors' narratives, which the neo-Lamarckian Romanes recognized. "There is not merely resemblance, but virtual identity, between the theories of stirp and germ-plasm," he wrote. "Disregarding certain speculative details, the coincidence is as complete as that between a die and its impress" (1899). Alfred Russel Wallace, on the other side of the heredity spectrum (as a hard-hereditarian), equally recognized that "the names of Galton and Weismann should therefore be associated as discoverers of what may be considered (if finally established) the most important contribution to the evolution theory since the appearance of the Origin of Species" (quoted in Churchill, 2015).[8]

Both Galton and Weismann based their new view of heredity on a series of antitheses – nature and nurture, innate and acquired, stirp and person for Galton, germ plasm and somatoplasm for Weismann. Nature, stirp, innate and germ plasm, the strong end of this dichotomous narrative, were internalized and reified, severed from external influences and made independent from the notion of individual generation. As Keller writes (2010), from that moment onward, "the alignment of the notion of inborn or innate with that of heredity" came to be seen as natural (the helpful comparison Keller makes with John Stuart Mill's pre-dichotomous writings well illustrates how peculiar and idiosyncratic was such an alignment and the oppositionality it implied).

Projecting forward slightly to topics we will later investigate in greater detail, hard heredity, by creating a strict division of inside and outside, also created new intellectual boundaries. On the one hand, it relegated the environment – the social – to an ancillary role in biology. On the other hand, biological marginalization was nurture's fortune: freed from the biological laws of heredity, nurture becomes a non-biological terrain, open to the exclusive observations of social science. The folding of the hereditarian substance in the germ plasm was turned into an excellent theoretical opportunity to emancipate the social sciences from biological heredity and reserved the social sciences' role to the mere study of "purely civilizational and non-organic causes" (Kroeber, 1917).

Thus hard heredity was an important conceptual rupture. Not only did it break with broader views of inheritance dominating medical and evolutionary thought in the nineteenth century, but its rise also had crucial implications for the making of the social sciences as a space distinct from the life sciences. By folding the biological into a germ plasm utterly separated from environmental inputs, hard heredity paved the way for a radical differentiation of the sociocultural from the organic.

However, before moving to the political-epistemological implications of hard heredity, we need to look at a third musketeer of the hard-hereditarian revolution, Wilhelm Johannsen, who finalized the architecture of hard heredity in the twentieth century, this time after Mendel's "rediscovery."

Johannsen and the "modern view of heredity"

Wilhelm Ludvig Johannsen (1857–1927), a largely self-trained Danish botanist who began his career as a chemist at the Carlsberg brewery later becoming a professor of plant physiology at the University of Copenhagen, wrote a nearly half-century after Galton's first writings on heredity, and two decades after Weismann's key contributions. His was a different epistemological context. Mendel had been rediscovered, and the word "genetics", firstly employed by the English geneticist William Bateson in 1906, was in scientific currency. Johannsen himself coined the term "gene" in 1909 "as a short and unprejudiced word for unit-factors" (1923), free, he claimed, of any theoretical hypothesis (1911).

Johannsen was explicitly a non-Darwinian – or, possibly, even an anti-Darwinian, as most Mendelians were at the time, something that may appear surprising to our modern synthesis' ears but that was not at the time.[9] Along with other Mendelians like Bateson, Johannsen did not believe that heredity could be reconciled with natural selection: the "Darwinian theory of selection," he claimed, "finds absolutely no support in genetics" (quoted in Roll-Hansen, 2009). Continuous variations – the stuff of Darwinians and Lamarckians – did not appear to fit neatly with genetics. Mendelians argued instead that discrete, discontinuous mutations were the key mechanism behind evolutionary novelty (see Provine, 2001; Gayon, 1998). This attitude was indicative of the rift between genetics and the theory of natural selection at the time, when genetic findings (Mendelism) seemed to disprove the continuous nature of variation implied by Darwinism (Provine, 2001; Bowler, 1983; Gayon, 1998; Depew and Weber, 2011). Mendelians were mutationists not Darwinian at the time. Biometricians in England objected vehemently to this attitude of Mendelians giving rise to one of the most significant controversies in the history of heredity. But also soft-hereditarians (neo-Lamarckians) perceived their work closer to Darwin than Mendelians did, an important fact to keep in mind when discussing debates on heredity and evolution in the first decades of the century.

Johannsen, anyway, had no sympathy for Darwinism. He saw Darwin's view of heredity via pangenesis as "primitive," a notion better suited to ancient times than to what he named the modern conception of heredity.

Definitely breaking with the past demanded for Johannsen a shift not only in the conception of heredity but in the disciplinary treatment of it. No longer could heredity be left to the descriptive, observational style of naturalists and morphologists who looked only at the visible characteristics of organisms. For someone trained in chemistry as Johannsen was, it was time instead to make heredity something "amenable to analysis just like the objects of chemistry" (Müller-Wille, 2007b). Johannsen's "'radical' ahistoric" view of heredity was "an analog to the chemical view", he wrote; chemical compounds have no compromising ante-act, H_20 is always H_20, and reacts always in the same manner, whatsoever may be the 'history' of its formation or the earlier states of its elements" (1911: 139). To advance his ahistoric view, Johannsen had to make an important distinction.

The genotype-phenotype distinction

Johannsen's main conceptual legacy lies in his introduction the genotype-phenotype distinction, a foundational step of classical genetics (Allen, 1978, 1979c; Roll-Hansen, 2009). In 1909, Johannsen put forth this new idea "as a consequence of the realization that the hereditary and developmental pathways were causally separate" (Lewontin, 2011). The suffix "type," common to both terms, clearly invites to a statistical (rather than individual) interpretation of heredity (Churchill, 1974), something that represents a difference from the parallel – and in a sense specular – opposition of germ plasm and somatoplasm in Weismann.

Genotype, Johannsen wrote, refers to the "the sum total of all the 'genes' in a gamete or in a zygote." Following the aforementioned chemical model, genotype has to be seen "as a complicated physico-chemical structure which reacts only as a consequence of its realized state, but not as a consequence of the history of its creation," (Sapp, 1987). Genotype indicates the "inner constitution" of an organism, but not the composition of that inner constitution. Johannsen wanted "genotype," as he did the term "gene," to be free of any theory, a "moratorium term" with regard to the proliferation of ids, pangenes and so on that characterized late-nineteenth-century cytological research on heredity (Churchill, 1974).

Where genotype refers to inner constitution, phenotype, from the Greek *phainein*, "to appear," refers to the visible traits of an organism. Johannsen defined phenotypes as:

> All "types" of organisms, distinguishable by direct inspection or only by finer methods of measuring or description ... the appearing (not only apparent) "types" or "sorts" of organisms are again and again the objects for scientific research. All typical phenomena in the organic world are *eo ipso* phenotypical, and the description of the myriads of phenotypes as to forms, structures, sizes, colors and other characters of the living organisms has been the chief aim of natural history, which was ever a science of essentially morphological-descriptive character.

In previous speculations, the concept of apparent, visible characteristics were used to describe links between generations. But, for Johannsen, a focus on the visible generated a misleading understanding of heredity. The transmission conception of heredity was in a sense a "phenotype-conception" of heredity (1911), based on the "superficial" view that what was transmitted were the visible characteristics of an organism. The most extreme version of the phenotype conception, Johannsen argued, was Lamarckian inheritance of acquired characteristics. This idea of heredity based on apparent levels of resemblance had to be rejected:

> The current popular definition of heredity as a certain degree of resemblance between parents and offspring, or, generally speaking, between ancestors and descendants, bears the stamp of the same conceptions [as the Lamarckian definition], and so do the modern "biometrical" definitions of heredity.... In all these cases we meet with the conception that the personal qualities of any individual organism are the true heritable elements or traits!

And he continued:

> The view of natural inheritance as realized by an act of transmission, viz., the transmission of the parent's (or ancestor's) personal qualities to the progeny, is the *most naive and oldest* conception of heredity. (1911, my italics)

In a move that clearly mimicked Galton's and Weismann's strategies of disassociating the personal qualities of an organism from the hereditary material, Johannsen, according to Roll-Hansen, claimed it was "possible to go beneath inheritance as the mere morphological similarity of parent and offspring and investigate the behaviour, and eventually the nature, of the underlying factors (genes) that were transmitted from one generation to the next" (Roll-Hansen, 2009).

Radically antagonistic to the phenotype or transmission view, the genotype conception, Johannsen argued in his "Genotype Conception of Heredity," meant that

> the *personal qualities* of any individual organism do not at all cause the qualities of its offspring; but the qualities of both ancestors and descendant are quite in the same manner determined by the nature of the "sexual substances" – i.e., the gametes – from which they have developed. Personal qualities are then *the reactions of the gametes* joining to form a zygote; but the nature of the gametes is not determined by the personal qualities of the parents or ancestors in question. This is the modern view of heredity.

A few pages later, Johannsen would brilliantly sum up this novel position with a famous formula: Heredity is "the presence of identical genes in ancestors and descendants" (1911). The domain of visible resemblance was swept away and replaced by a continuity at the genotypic level very much in sync with Galton's stirp and Weismann's germ plasm.

Johannsen described his work as "initiated by Galton and Weismann, but now revised as an expression of the insight won by pure line breeding and Mendelism." This alludes to one of Johannsen's key experiments, carried out with self-fertilizing, or pure-line, beans. The experiments demonstrated that no matter the different environmental inputs, within each pure line, variation was minimal. Phenotypic effects did not seem to matter at all at the genotypic level: the degree of hereditary stability took even Johannsen by surprise (Roll-Hansen, 2005). The message was that "the personal character of the mother-bean has no influence, that of the grandmother, etc., also none; but the *type of the line* determines the average character" (Churchill, 1974). Pure-line experiments reinforced decisively the "hardness of genotype and genes," Roll-Hansen writes. "Heredity is generally stable and changes only intermittently. There is no continuous change of heredity as assumed in orthodox Darwinian or neo-Lamarckian theories" (Roll-Hansen, 2009).

The notion of pure line was taken up in the harsh political debates of the following decades: Soviet Lamarckians, in particular, saw in it proof of the intrinsically fascist nature of Mendelism. However, Johannsen's ahistoric view of heredity left no explicit room for such overtones. It is also fair to say that Johannsen's view was not that of a simplistic Mendelian: Johannsen "never accepted an interpretation of genotype which simply made it identical to chromosomes or any other physico-chemical structures or particles" (Roll-Hansen, 2009).

Johannsen was cautious about the parallelism between his genotype and phenotype and Galton's nature and nurture. He said that the phenotype had to be seen as the result of an interaction between genotype, nature, and nurture. But the history of hard heredity is nonetheless marked by a succession of antitheses, albeit not perfectly aligned. Nature (Galton), germ plasm (Weismann), and genotype (Johannsen) became increasingly conflated showing a biological reality progressively identified with something running inside ourselves, the genetic. Nurture, the soma, and the phenotype, do not map perfectly onto one another either, but in a sense they merged in the pole of what is ephemeral, opposed (and even caused) by the genetic level. A final word on the use of the term "modern" employed by Johannsen to describe his conception of heredity. If expanding on Latour (1993), we take modernism as a strategy of purification, dichotomization and elimination of hybrids, what Galton, Weismann and Johannsen achieved was exactly a modernization of the view of heredity. Immunized from any dangerous confusion with the body or sociocultural factors, free from any unnecessary admixture of nature and nurture, the modernist architecture of heredity was clear, neat and functional, the perfect penchant for the idea of an autonomous, self-contained individual looking for an exact correspondence between personal natural talent (later genetic endowment) and place in society.[10]

There is, finally, a fourth significant moment in the modernization and hardening of heredity: its translation to the molecular level with Francis Crick's dogma of molecular biology in 1958, which we will turn to briefly in Chapter 5 and then again in 7. For now, let us see how the penetration of modern view of heredity into the public sphere was to have important political and epistemological implications: at the level of politics, in attracting attention to human heredity as a well-defined, self-contained space of management and control; and in terms of knowledge production, in reconfiguring the border between life and social sciences.

3
Into the Wild: The Radical Ethos of Eugenics

The originality of the eugenic experience: 1900–1945

The making of the modern (hard) view of heredity ushered in a dramatic conceptual transformation. This meant, in detail, the creation of a space of nature as opposed to nurture, of the stirp as opposed to the ephemeral individual, of the immortal germ-plasm as opposed to the transient reality of the body. These new discourses gave rise to a series of conceptual antitheses – innate/acquired, heredity/development, germ-cells/somatic cells, unchanging hereditary material/transient lifetime experiences, and later genotype/phenotype – that were not there in nineteenth century developmental biology, and would supply the conceptual arsenal on which much of twentieth-century politics and epistemology of the biology/society border was constructed.

The making of hard heredity was the result of specific sociopolitical and ideological pressures, particularly evident in Galton. However, if various political and ideological motives, alongside scientific and technological ones, contributed to the rise of hard heredity, how was this epistemic construct to be further politicized "under the existing condition of law and sentiment" as Galton put it (1909)? Or, to rephrase in more sociological terms, how, in the context of Western demographic, economic, and cultural trends between late nineteenth and early twentieth century, was the making of this new view of heredity to enter and shape the political debate?

It would do so mainly via eugenics, which treated the insulated hereditarian space paradoxically. On the one hand, as the first eugenicists said, heredity offered "a biological source over which" human beings have "no control" (Crackanthorpe, 1909). On the other hand, humans must take control of heredity for the sake of civilization and the achievement of social and economic goals. This dilemma was at the core of the politics of the eugenic movement in its multifaceted version.

Eugenics is not synonymous with a specific theory of heredity. Eugenics could be soft hereditarian as well, depending on national and cultural contexts. And even when eugenics was hard hereditarian, it could take the most disparate political forms, from the far right to the far left, from Hitler to some left Mendelian Bolsheviks. But although it is wrong to conflate hard heredity and eugenics, it is also wrong to overlook the important role that hard heredity played in the making of the eugenic movement.

While hard heredity did not directly cause eugenics, without it, the politics of human heredity would have lacked its pulsing heart. Hard heredity made the core of the human investigable, quantifiable and manageable. In the absence of hard heredity, debates on population quality would have lacked coherence. They would have remained diluted by medical, environmental, social reform discussions, as in the nineteenth century. Hard heredity thus functioned as an ideological catalyst of the international debate, even when it was opposed by soft-hereditarian trends within the eugenic movement. After the rise of hard heredity, it was no longer possible to fix the problem of bad heredity merely via social reform, as it was instead under a developmentalist and soft-hereditarian picture (Paul, 2006).

Alexander Carr-Saunders, a distinguished demographer and a young member of the British Eugenics Society (and future successor of William Beveridge as LSE director), clearly recognized this dependence of the politics of eugenics on a certain epistemological view of heredity:

> Eugenics, as generally understood, is based upon the assumption that it is to the germ-plasm and not to the environment that we must look, when we seek the principal agent which determines the characteristics of future generations. If acquired characters are to any extent inherited, then to that extent we are thrown back upon the environment. (Carr-Saunders, 1913: 217)

Similarly, in *The Problem of Race-Regeneration* (1911), Henry Havelock Ellis recognized the importance of nineteenth-century social reform movements but noticed that they "touched only the conditions of life and not life itself":

> We have been expending enormous enthusiasm, labour, and money in improving the conditions of life, with the notion in our heads that we should thereby be improving life itself, and after seventy years we find no convincing proof that the quality of our people is one whit

> the better than it was when for a large part they lived in filth, were ravaged by disease, bred at random, soaked themselves in alcohol, and took no thought for the morrow.

For Havelock Ellis, eugenics was the movement for the "betterment of life itself," and hard heredity delimited what one could call such a space of life itself. On this reified notion of life and heredity, the huge and pluralistic debate on the politics of human heredity took place.

However, if hard heredity was the attractive force that precipitated the emergence of the eugenic problematics but not its unifying mechanism, what did unite the international eugenics movement? My argument in this chapter is that the varieties of eugenics in the first decades of the twentieth century were united less by direct politicization of a particular theory of heredity than by *a common ethos*, which crossed over the linguistic and aesthetic borders dividing eugenic approaches.

This ethos, or moral economy, is recognizable by four main features that were extremely visible between 1900 and 1940, common to all the components of the eugenic movement (from right to left, from hard hereditarians to Lamarckians) and substantially disappeared after that.

First, radical biologism. Eugenics flattened the notion of the human and its psycho-cultural manifestations into its merely biological dimension. Even mental and moral qualities were seen as aspects of the hereditary mechanism. Galton's most influential pupil, statistician and biometrician Karl Pearson, put it this way:

> We are forced...to the general conclusion that the physical and psychical characters in man are inherited within broad lines in the same manner, and with the same intensity. (Pearson, 1903: 204)

Someone like Austrian neo-Lamarckian Paul Kammerer (at the opposite of Pearson on many points, as we shall see in a next chapter), would disagree a lot from Pearson on *how* these mental qualities could be inherited but not that *they could be* passed on through generations as other physical characteristics of human beings.

Second, utopian social engineering. Eugenicists believed that the future of human evolution had to be actively controlled by human efforts. They emphasized social engineering, often in ameliorist and even utopian terms, as we shall see. All of them believed in a management style of political intervention similar to the authoritarian "high modernism" James C. Scott describes in *Seeing Like a State*, an ideology of "rational design of social order commensurate with the scientific understanding of natural laws" (1998). In particular, they were confident in science as the basis for the rational planning of nature and society.

This brings us to the third axis of the eugenic ethos, the unlimited empowerment of scientific experts. Eugenicists saw the sciences as appropriate, morally neutral additions to political and ethical debates. They thus saw a prominent public role for scientific experts. Science and policy were for eugenicists in a relationship of immediate interchange. The key concern for them was one of dissemination of the eugenic gospel, downward toward the wider society (Larson, 1991). After all, these uncontested scientific authorities saw themselves as holding the key to controlling human reproduction.

Finally, primacy of the race over the individual. Eugenics subordinated the good of the individual to the health of the species or race. Eugenicists believed that an individual act of reproduction was not private but a social instrument by which the betterment of the race could be guaranteed. Therefore the subjective preferences of the individual were irrelevant. Charles Davenport, a leader of American Eugenics, put it bluntly. The happiness of each parent, he said, had "little eugenic significance" (1911).

As I will argue in Chapter 5, all of these features were contested and eventually undermined by broader changes in the moral economy of post-world war II Western societies and, more specifically for my argument, they became increasingly incompatible with the new thought-style of political biology that the architects of the modern synthesis, in particular, promoted since the late 1930s. This is why it makes sense to demarcate neatly a well-defined era in the politicization of human heredity beginning ca. 1900[1] and ending sometime between the late 1930s and 1945, depending on the country. Ending, it is worth repeating once again, does not mean that eugenic measures like coerced sterilization were discontinued overnight after 1945. They lasted in many important countries, like the U.S., until the 1970s, when the effects of civil rights and anti-authoritarian political movements undermined their legitimacy (Largent, 2008). Nor does ending mean that the legacy of eugenics was not extremely significant in many medical post-war debates (Bashford, 2010). What ended, as I will explain next, is a specific style of politicization of biology, marked by the four axes I have listed above.

The many meanings and ideologies of eugenics

Political concerns for the quality of the human stock and the social control of human reproduction can be traced back to Plato's *Republic*, continuing into early modernity with utopian writers such as Thomas More and Tommaso Campanella (Hertzler, 1923). In the Book IV of *Republic*, "Socrates explicitly compares the breeding of humans to that of animals", and in Book V "he argues that we must 'mate the best of our men with the best of our women as often as possible, and the inferior men with the inferior women as seldom as possible, and bring up only the offspring of the best'" (Parrinder, 1997; Plato, 1987).

However, only when the explicit preoccupation with biological heredity (and with a well-demarcated, reified, notion of it) emerged in the second half of the nineteenth century did a science and a politics aimed at the systematic improvement of the human stock come overtly to circulate in society. This is where Galton, the polymath with a scientific interest in the distribution of natural talent, overlaps with Galton the political theorist of heredity.

Though Galton's ideas were not initially accepted in the mainstream, many professional groups – biologists, doctors, educators, social reformers, and even novelists – did eventually find themselves attracted to the emerging international eugenics movement. The most significant cultural change that favored Galton's ideas was the shift from the social optimism of the Victorian period to a much more negative attitude toward the potentialities of social change. As historian of anthropology George Stocking wrote, eugenics

> was the product of a period when traditional liberalism, threatened by forces of democracy and collectivism at home, and by those of nationalism and militarism abroad, was no longer the optimistic creed it had once been. Rather than a positive application of the principles of natural selection to the interpretation of social phenomena, eugenics was an attempt to compensate for the failings of natural selection to operate under the social conditions of advanced civilisation. (1987:145)

Many factors were at play in this transition from the Victorian spirit of optimism and individualism – of which social Darwinism was the most direct political representative – to the very different cultural terrain in which eugenics flourished. In Britain, where eugenic speculations emerged first, demographic fears were paramount, especially the differential fertility between upper and lower classes, a "new population question" to which various strands of British eugenics, from Fabians to biometricians, were particularly sensitive (Soloway, 1990). Anxiety of degeneration and "national efficiency" were everywhere, apparently confirmed by the events of the Boer War (Searle, 1971, 1976; Soloway, 1990). Several decades later Richard Titmuss would characterize these fears as "astonishingly gloomy prophecies" (1944) of the superior classes outbred by the inferior (a theme originating with Darwin and Galton).[2] Alongside were economic anxieties driven by urbanization and industrialism that merged with a "taxpayer revolt against the burdens of pauperism, crime, and insanity" (Crook, 1994).

The trend was not unique to Britain. Throughout Europe, crowds, immigrants, cities, and modernity itself gave rise to concerns about national decay and degeneration. Such anxieties reflected "a counter-theory to mass-democracy and socialism" (Pick, 1989; see also Carlson, 2001) long before the rise of eugenics. Authors as diverse as Bénédict Morel, Max Nordau, Émile Zola, Henrik Ibsen, Cesare Lombroso, Ray Lankester, Hippolyte Taine, Friedrich Nietzsche, and Gustave Le Bon worried about a decline in the quality of the human stock (Pick, 1989). In his 1901 article "The Causes of Racial Superiority," the American sociologist Edward Alsworth Ross, coined the expression "race suicide," condensing these anxieties about inferior races and classes out-breeding their betters.

These anxieties were the daily bread of eugenicist thinkers (and certainly also in the originary repertoire of Charles Darwin himself who overtly expressed his concern for how the weakest members of society outbreed the rest). In the first issue of *Eugenics Review*, the official journal of the British movement, William R. Inge, dean of St Paul's Cathedral and a Cambridge professor of divinity, voiced similar dysgenic fears of "reversion" in the quality of the human stock:

> One general principle which I believe to be incontestable is, that if natural selection is inhibited, if nature is not allowed to take her own way of eliminating her failures, rational selection must take its place. Otherwise nothing can prevent the race from reverting to an inferior type. (1909: 29)

It is in the context of this "crescendo of interest in what became known as the dying or doomed races" (Levine, 2010) that a break with the optimism and "facile environmentalism" of Victorian reformers became increasingly appealing (Searle, 1976; Haller, 1963; Soloway, 1990).

The laissez faire individualism dominant when Galton first wrote on heredity also became outmoded as the liberalism of the late nineteenth century showed itself inadequate to solve the economic and demographic problems of industrialization and capitalism (Allen,

1987). Planning and social intervention were no longer bad words in the post-Victorian scenario. In the United Kingdom, Sidney Webb, one of the leaders of Fabianism, captured the new attitude: "No consistent eugenicist can be a 'Laisser Faire' individualist unless he throws up the game in despair. He must interfere, interfere, interfere!" Frank Fetter, an influential American economist, voiced similar concerns. "Unless effective means are found to check the degeneration of the race," he said, "the noontide of humanity's greatness is nigh, if not already passed. Our optimism must be based not upon *laissez faire*, but upon vigorous application of science, humanity, and legislative art to the solution of the problem" (both quoted in Leonard, 2005).

Meanwhile in Germany, medical humanitarian values were in decline, and the racial hygiene movement was giving birth to an "authoritarian collectivism" driven by state-oriented experts. This novel ethics of race was especially attentive to population policies. The good of the people was no longer to be left in the hands of individual philanthropists (Weindling, 1989).

This abandonment of laissez faire policies reflected the transition from nineteenth-century liberalism to a "new philosophy of regulated capitalism." Eugenics, with its "regulation of human reproduction itself," was a perfect "biological counterpart" from the standpoint of the ruling classes (Allen, 1987: 182). In an age of increasing emphasis on social control amid anxieties about decadence and degeneration, Galton's definition of eugenics as the "study of *agencies, under social control*, which may improve or impair the racial qualities of future generations either physically or mentally" (Galton, 1884; my italics) seemed finally perfectly suitable.

Varieties of eugenics

Eugenics is "the science which deals with all influences that improve the inborn qualities of a race" (Galton, 1904). But within this, there are distinctions. An important one, advanced (as he claimed) by the British eugenicist Caleb Saleeby, is that between *positive* and *negative* eugenics.

Positive eugenics refers to the improvement of the human race by supporting the reproduction of the fittest members of society, though just who was fittest varied significantly depending on the social and political agenda of the eugenicist. The model might range from the successful capitalist to the successful revolutionary, such as Vladimir Lenin (as H. J. Muller wanted). The traits proposed as sources of success spanned altruism and strength, leadership and height, intelligence and luck (Adams, 2009: 379).

Positive eugenics was not confined to socially progressive authors. "The investigation of human eugenics," Galton wrote, concerns "the conditions under which men of *a high type* are produced" (1883; my italics). Harry Laughlin, a right-wing American eugenicist, emphasized the "art of breeding better men" in order to secure "high fertility and fittest mating among the more talented families" (Laughlin, 1919).

Details of positive measures, particularly visible in utopian writings of leftist eugenicists such as J. B. S. Haldane (1924), Herbert Brewer (1935), and H. J. Muller (1936), included not only taxation and education but also more radically "ectogenesis," the equivalent of *in vitro* fertilization and development of human eggs, "*eutelegenesis*" (artificial insemination), and the establishment of what we would today call sperm banks.

However, positive eugenics remained mostly a theoretical idea, with no practical applications. This was partly due to technological deficiencies. Another obstacle was ethical quandaries, which were deemed to be a special problem of positive eugenics (Paul 1988; Adams, 2009), rather than, as we would tend to believe with our current standard, the practices of negative eugenics.

Negative eugenics, conversely, was widely applied in Western society. Such measures aimed to prevent the diffusion of what was then called "defective germ-plasm." In Laughlin's definition, the goal of these restrictive policies was to cut off

> the descent-lines of those individuals who are so meagerly or defectively endowed by nature that their offspring are bound to be antisocial, or at least unable to care for themselves. (1909)

In practice, negative eugenics mostly meant forced sterilization (which for men meant vasectomy, and for women tubal ligation). The first sterilization law in the United States was passed in Indiana in 1907 in light of concerns about miscegenation and racial degeneracy. Similar worries prompted the 1924 Immigration Act and the Supreme Court decision in *Buck v. Bell* (1927), which upheld Virginia's forced sterilization laws. In the early 1930s, twenty-nine states would pass similar laws, sterilizing approximately 30,000 "feebleminded" people, a figure that rose to 63,000 by 1960s, and is probably an underestimation (Largent, 2008; see also Stern, 2005).

The first European sterilization law was enacted in Vaud Canton, Switzerland in 1928, followed swiftly by similar laws in the Scandinavian countries and Germany, but also most of Eastern Europe, Cuba, Turkey and Japan (Paul, 2006). Some of these laws were discontinued only in the 1970s. Under Nazism the effect of negative measures escalated to an unprecedented level, culminating in the infamous Action T4 euthanasia program (which actually and more crudely meant gassing or lethal injection), itself a grim prelude to the industrial extermination of the Holocaust (Weindling, 1989; Proctor, 1988).

What joins negative and positive eugenics is the construal of reproduction as a political problem that could not be solved by individuals. Given the often-irrational form taken by reproduction in human societies, the process had to be controlled and rationalized so to avoid undoing the work of natural selection.

Eugenicists cited many reasons for their effort. For one, they perceived social life as unnatural; the regulated order produced by natural selection was threatened by measures of social benevolence and tolerance that would eventually destroy the nation. Articles in *The Eugenics Review* are replete with sad considerations about the fall of civilizations, the disappearance of cultures and great empires. This is one of the clearest eugenics concerns, stemming directly from Darwin himself: Civilization led to a relaxation in the operation of natural selection, and this had to be fixed (Bajema, 1976). In the words of Leonard Darwin, son of Charles and a leader of British eugenics, the task was "to substitute for the slow and cruel methods of nature some more rational, humane, and rapid system of selection by which to ensure the continued progress of the race" (1912). Indeed, it was an unacceptable scandal that human reproduction was left to mere chance when so much care was given by breeders to select cattle or plants – a point made by Plato more than 2,000 years earlier as early eugenicists were ready to notice. The professional knowledge of animal and plant breeders thus played a key role in the effort to improve human stock and fitness.

There was also a very peculiar aesthetics and visualization of eugenics (Mazumdar, 1992; see Stillwell, 2012). Eugenics became known to the public via pedigrees, tables and diagrams for biometricians, meticulous charts of recessive and dominant traits for Mendelians. Their function was to make visible to the wider society the iron law "of the basic hereditarian claims that like produces like" (Mazumdar, 1992). Many of these genealogical diagrams of families were increasingly to be found in medical journals (Levine and Bashford, 2010).This interest in aesthetics of course pervaded the eugenic body. Through measurement to confirm the inferiority of certain races, eugenicists visually pathologized the unfit. (Stillwell, 2012). More generally, as historian Nancy Leys Stepan has written, the aesthetic sensibility of eugenicists was dominated by a concern "with beauty and ugliness, purity and contamination" and by an obsessive search for correspondence between external markers of fitness and inner genetic worth (Stepan, 1991: 135).

However, the emphasis on the visible aspects of bodily fitness should not cloud what is probably the key feature of the philosophy of eugenics: in human beings, heredity governs the visible as well as the invisible, the physical as well as the mental, eye color as well as character and personality. Galton wrote as much in 1869. "A man's natural abilities," he averred, "are derived by inheritance, under exactly the same limitations as are the form and physical features of the whole organic world."

Writing in the 1920s, R. A. Fisher could hardly believe that early researchers, such as Carr-Saunders and Alfred Russel Wallace, had criticized the "firmly established" notion that "the mental and moral qualities of mankind are inherited to the same extent as are the physical characters" (Fisher, 1922a: 190). As Fisher made clear in his seminal *The Genetical Theory of Natural Selection*:

> Man, like the other animals, owed his origin to an evolutionary process governed by natural law; and next that those mental and moral qualities most peculiar to mankind were analogous, in their nature, to the mental and moral qualities of animals; and in their mode of inheritance, to the characters of the human and animal body. (1930: 170)

The claim of equal ontological status between the physical and moral is far from new. To limit ourselves to modernity, that position was also found in the radical Enlightenment tradition that, since Pierre Jean George Cabanis, denied any special ontological status to *le moral* over *le physique*. However, in the post-Darwinian moment, the notion took on a different significance. Human nature became a dynamic reality subject to change, with an evolutionary past and therefore various possible evolutionary futures. In an article significantly titled "Some Hopes of a Eugenist," Fisher argued that Darwin's theory was not to be seen merely as "a description of the past, or an explanation of the present" but the "veritable key of the future." Eugenicists believed that by exercising vigilance – by grasping the logics and trends in human populations – they would corral humanity toward the best possible versions of that future, utopian societies regulated solely by improvement of the genetic stock.

A revision of eugenics

The historiography on eugenics has significantly broadened since 1990. New scholarship approaches eugenics as a "polysemic" (Koch,

2004) system of thought, endowed with a "tantalizing ethical complexity" (Adams, 2009). Eugenics, says Marius Turda, is "a cluster of diverse biological, cultural and religious ideas and practices that interacted with a variety of social, cultural, political and national contexts" (Turda, 2010). Far from being a monolithic phenomenon, eugenics is increasingly recognized as a "broad church" encompassing a vast spectrum of epistemic and political views, a "very mixed bag" (Cowan, 2008). Overall, we can ascertain six shifts in the historiography.

First, it has become evident that eugenics was an international movement with different national contexts, agendas, and scientific presuppositions (Adams, 1990; Stepan, 1991; Allen, 2011). The Anglo-American and German experiences are no longer seen as universal cases. As Mark B. Adams reminds us, "By 1935, eugenics and other related ideas had spread widely, and eugenics societies and movements had been founded in more than 40 countries," from Cuba to Japan, from Europe to Brazil, from North America to Australia (Adams, 2009; Bashford and Levine, 2010).

Second, eugenics was epistemically pluralistic in theories of heredity. Plenty of national eugenics movements, especially in Latin countries that were not significantly touched by Mendelism, are proof of this epistemic complexity as well as the "ubiquity and even normality of eugenic themes" (Leys Stepan, 1991:5). As William Schneider writes:

> The French experience shows that a Mendelian hereditary theory was not a necessary condition for the development of eugenic thought. Mendelism did not come to France until the 1930s, yet from the beginning of the century Lamarckian hereditary theory, which maintained that acquired characters *could* be inherited, was the basis of a eugenics movement with similar goals and some of the same programs as those in Anglo-Saxon countries. (Schneider, 1990: 217 and ff)

It is important to keep in mind that there was not only a eugenics tradition in Latin countries, where the issue was not "of differential fertility but of the generally low fertility" (Mazumdar, 1992). Even in northern nations, where Mendelism rapidly took hold, important Lamarckian figures still countered the Mendelian approach advocating eugenic policies. This is further evidence, as will be seen in next chapter, that while hard heredity undergirded the politics of human heredity, it did not determine the outcome of debates surrounding those politics.

Third, though eugenics is popularly associated with reactionary political trends, it in fact flourished equally under liberal, fascist, and socialist governments (Mottier, 2010; Bashford and Levine, 2010). Eugenics, in Haldane's observation (1938, quoted in Paul: 1984), "cut right across the usual political divisions."

Racial hygiene in Germany developed under the social democratic Weimar Republic as well as the Third Reich (Weindling, 1989). From the Fabians to members of the German Social Democratic Party, from French Communists to Soviet Bolsheviks, from health and birth control activists to feminists, from Harold Lasky and Kammerer to Muller, Haldane, and the Myrdals, eugenics was part of the culture of the left in its various political and scientific styles. As Diane Paul writes in *The Politics of Heredity* (1998), "Virtually all of the Left geneticists whose views were formed in the first three decades of the century died believing in a link between biological and social progress". Even John Maynard Keynes endorsed eugenics. In 1946 he called Galton "the founder of the most important, significant and, I would add, *genuine* branch of sociology which exists, namely eugenics." Articles in *Eugenics Review* praised socialism as a natural and powerful ally of the eugenics movement (Herbert, 1910).

This is not to deny that many leftist authors – including John Dewey, Franz Boas, and Lancelot Hogben – were profound critics of eugenics. But opposition to eugenics, too, cut across political lines. Conservative authors, such as Chesterton and Pope Pius XI, were also harsh critics of eugenics. Mistrust of experts was one key conservative concern about eugenics.

Fourth, and relatedly, recent scholarship tends to understand eugenics as a phenomenon in itself, not just through the prism of Nazism (Turda, 2010) or as the forerunner of the Holocaust (Taguieff, 1991; Weiss, 1990). No one is thereby acquitted by history, but we should understand that eugenics became a dirty word mostly after Nazism took power. It would be wrong to deny that eugenics was an "integral aspect of European modernity" (Turda, 2010; Weindling, 1989; see also Griffin, 2007), aimed at transforming the nation "into an object of scientific regulation and expertise" (Turda, 2010). Eugenics was part of a "larger biopolitical agenda that included social and racial hygiene, public health and family planning as well as racial research into social and ethnic minorities.... both advocated and adhered to by professional and political elites" (Turda, 2010:2). As Frank Dikötter (1998) suggests, "Eugenics was not so much a clear set of scientific principles as a 'modern' way of talking about social problems in biologizing terms" (see also Jackson, 2005). Whereas historians once focused almost exclusively on the scientific dimension of the eugenics movement, today historians focus as well on the links among eugenics, nationalism, and nation building in the early twentieth century (Turda, 2010; Bashford and Levine, 2010).

Fifth, eugenicist and environmentalist strategies have been increasingly conceptualized as parallel and even complimentary strategies of population management (MacKenzie, 1982; Werskey, 1979; Rose, 1979; Freeden, 1979). Kathy Cooke, for instance, has emphasized that, at least until 1915, no clear division between heredity and environment was evident in American eugenics. Only later did hard heredity marginalize soft eugenics approaches (1998). *Euthenics*, from the Greek for "prosperity," "fortune," "abundance," reflects this comingling of eugenics and environment. According to Ellen Richards – a pioneering woman in science home economics and public health – euthenics was meant to convey the idea of "race improvement through environment." This was not in opposition to eugenics but was, rather, a "preliminary science on which Eugenics must be based" (1910). Its main targets were the improvement of the environment through "sanitary science," "education," and "relating science and education to life." Saleeby's nurtural eugenics in Britain was based on similar principles. We shall discuss this at length in the next chapter.

Finally, the conventional view that eugenics (or racial hygiene) was a discredited guest in respected genetics circles, especially after the 1920s (Ludmerer, 1972; Rosenberg, 1967), and that it was based on a poor understanding of genetics (see for instance, Haller, 1963, Bajema, 1976; Carlson, 1987 Cravens, 1978), has been challenged over the last three decades by a second wave of historical studies (Paul, 1995, 1998; Barker, 1989; Kühl, 1994). It is true that geneticists such as Morgan, who privately withdrew support, and Herbert Spencer Jennings, at Johns Hopkins (see Allen, 2011), considered much of eugenics second- or third-rate science. However, sophisticated scientists such as Fisher and Muller were passionate supporters of eugenics and for contrasting political reasons. In both cases eugenics profoundly informed their scientific view. We shall discuss Muller at a greater length in Chapters 4 and 6. As for Fisher, the last third of his hugely influential *Genetical Theory of Natural Selection* (1930) which he called "strictly inseparable from the more general chapters", displayed the full eugenicist repertoire – decay of the ruling classes, differential birth-rate, and so on – with the goal to shape "the practical affairs of mankind" (275). His view of genetics also incorporated eugenicist ideas such as "progress through selection" (Norton, 1983; see also: Mackenzie, 1982; Kevles, 1985; Soloway, 1990; Plutynski, 2006; Tabery, 2008).

Historians who has studied the attitude of geneticists with regard to Mendelian theories of mental defect, claims that, in spite of scientific shortcomings, many geneticists didn't abandon eugenics at all:

> Indeed, most welcomed it with open arms, impervious to the criticisms which were mounted during the 1920s.... On this evidence abandonment and repudiation was less complete, less consistent, less rational, and less scientifically grounded than has been suggested; it also perhaps came later. (Barker, 1989:375)

In a similar fashion, in *The Nazi Connection* (1994; see also 2013), Stefan Kühl contests the idea that after the 1920s geneticists and eugenicists no longer communicated, thereby challenging the notion that new discoveries in genetics undermined the core of eugenics doctrine. He also shows that links between American and Nazi eugenics were severed by World War II, not by intellectual transformation (see also Weindling, 2004). Furthermore, Daniel Kevles's division between mainline (Laughlin, Davenport, Steggerda, Grant) and reform eugenics (Hogben, Haldane, Penrose, Jennings, Conklin) doesn't imply that the latter broke with core eugenics doctrines. In Diane Paul's words:

> Nearly all geneticists of the 1920s and 1930s – including those traditionally characterized as opponents of eugenics – took for granted that the "feebleminded" should be prevented from breeding. (1995: 70)

A major example of this attitude is in 1939 The *Geneticists' Manifesto* (Social Biology and Population Improvement, promoted by Muller and signed by all the major geneticists, Fisher excluded, most of them "reform eugenicists") published at the Genetics conference in Edinburgh, and still very much rooted in a eugenic worldview (in progressive sauce, an example of that Left-Mendelism we shall explore soon) with overt support for voluntary (temporary or permanent) sterilization. Despite its dubious epistemic premises, eugenics lost much of its appeal not because it was proven mistaken but as a result of changing political values; as Diane Paul and Hamish Spencer famously claimed (discussing the impact of the Hardy-Weinberg principle on the eugenics movement): "there was nothing wrong with most eugenicists' math. Our concept of right, however, is much more expansive than theirs" (Paul, 1998). However, some of these shifting values and broader concept of right had an impact on the "math" (meaning technicalities) of evolutionary thought since 1940s, or at least helped bringing to light and consolidating some conceptual orientations that were much more resonant with the mutated scenario, as we shall explore in Chapter 5.

Class and race in Anglo-American eugenics

The particular political anxieties to which eugenics responded reflected the national circumstances in which the movements took place. The two most studied branches of the movement, the British and American experiences, are cases in point. The way these two movements were eroded in their credibility is well known. In the United States, eugenics lost steam mostly because the Great Depression affected every stratum of society: both "fit" and "unfit" were laid lower by economic crisis (Stillwell, 2012). In Britain eugenics became increasingly unacceptable thanks to its association with Nazism. However, what were the key tenets of these two movements?

Mainstream historiography has highlighted the class element of British eugenics: the target was mostly pauperism and the urban poor, (Searle, 1976, 1979; Mazumdar, 1992) the vast *lumpenproletariat*, especially of London, the "unskilled residuum" or "the army of the biological unfit" as eugenicists called them (Searle, 1981). In the British movement, proposers of alternative form of eugenics were aware of this classist bias of the Eugenics Society and accused "'class prejudice' to be the worst of all the enemies that eugenics has to face" (Saleeby, 1914). It is, however, interesting to note that as a "professional middle class" phenomenon (Searle, 1976; MacKenzie, 1979, 1982; Mazumdar, 1992; see for a different view Jones, 1986) eugenic criticisms were at times marshaled against "our present sham nobility (...) that means nothing biologically" as well, on the grounds of a meritocratic ethos (Schiller, quoted in Soloway, 1990; see also Galton, 1869 from which the argument originates).

The obsession with paupers meant for British eugenics the urgency to address the problem of "fertility in relation to economic status." This anxiety was expressed in the key formula of differential birthrate: the lower classes, i.e., "the less endowed," were breeding at a much faster pace than the "wealthy stocks." As Pearson wrote, "We have two groups in the community, one parasitic on the other. The latter thinks of to-morrow and is childless, the former takes no thought and multiplies" (Pearson, 1905). To discern this target, in 1906 David Heron, a researcher in the Francis Galton Laboratory for National Eugenics, created a highly systematic study of the city of London over a period of fifty years.

As late as 1930, Fisher, one of the fathers of the modern synthesis, was complaining that the "birth-rate is much higher in the poorer than in the more prosperous classes, and that this difference has been increasing in recent generations." Fisher warned of "disastrous biological consequences to which our own [civilization] seems exposed." In response to this perceived crisis, he proposed a scheme of family allowances to prevent the lower classes from out-breeding the middle and upper classes (1930: 245).

Needless to say, these anxieties of a decay of "the racial qualities of future generations" (Fisher, 1926: 116) were easily connected to a broader taxpayer resentment toward the increasing financial burden of the welfare system (Soloway, 1990: 90; Crook, 1994), an argument present from the very beginning of the eugenics saga. That eugenics inspired calls for a various way of reforming the welfare state is hardly surprising. After all, eugenics can be seen as an approach to political economy, the attempt to rationalize the use of human-biological capital (Turda and Weindling, 2007). As Galton made clear, if eugenics distinguished people's worth in their childhood, "it would be a cheap bargain for the nation to buy them at the rate of many hundred" (1901).

Histories of British eugenics have generally focused on a middle-class movement marked by class prejudice and a desire for economic rationalization. But there was more to it than that. British eugenicists, mindful of the intermixing brought about by empire, were fearful of race crossing and "degeneracy" as well as the increasing presence of immigrants on the Isles themselves. Galton, Pearson, Leonard Darwin, and the Lamarckian Ernest MacBride all emphasized what they saw as corruptions of pure stock (on Galton see Yudell, 2014; also Cowan, 1985; more generally Stone, 2001, 2002). This focus on race is helpful to correct the distinction between a supposedly non-racialist British eugenics versus the racist version of eugenics and race hygiene in America and Germany. "Even if not yet having acquired the biologistic hue that Nazi eugenics would later take on," Dan Stone writes, "eugenics in Britain was, on both the left and the right, a basically racist enterprise" (2001; on Fisher's ambivalent attitude toward race mixing included an attempt to rescue Gobineau's view see, 1930:257 and ff).

Still, the racism of British eugenics was muted in comparison to that of its American counterpart. American eugenics (Haller, 1963; Pickens, 1968; Ludmerer, 1972; Allen, 1976, 1979a, b, 2002; Kevles, 1985; Reilly, 1991; Selden, 1999; Carlson, 2001; Jackson, 2005; Bruinius, 2007; Largent, 2007; Lombardo, 2008; Yudell, 2014) was fundamentally right wing and had significant connections with Nazism (Kühl, 1994, 2013; Black, 2003). While the urban poor loomed large in the imaginations of British eugenicists, the specters of the immigrant and the feebleminded, and sometimes a combination of the two, were foremost on the minds of American eugenicists.

From Madison Grant's *The Passing of the Great Race* (1916) to Paul Popenoe and Roswell Hill Johnson's *Applied Eugenics* (1918), American eugenics publications were full of anti-immigrant fervor, anti-black racism, anxieties of race degeneration and debilitation, and fantasies of *mongrelization* and race purity. Even the more technical publications, such Davenport and Steggerda's study of race crossing in Jamaica, asserted fundamental differences in mental capacity between blacks and Europeans. After discussing the "variability of each race and sex in respect to each bodily dimension and many bodily organs," Davenport and Steggerda concluded there was a firm link between morphological and mental traits: whites were superior to blacks intellectually, while blacks were better than whites at sensory tasks and had a stronger sense of rhythm. A portion of the mixed-race sample was "mentally inferior to the Blacks." Hybridization invariably produced disharmonies, especially "in the mental sphere." The message was unambiguous:

> It seems to us the outcome of the present studies is so clear as to warrant the conclusion that they put the burden of proof on the shoulders of those who would deny fundamental differences, on the average, in the mental capacities of Gold Coast negroes and Europeans. (1929)

Years before, Davenport had also helped to define feeblemindedness, as:

> the acts of taking and keeping loose articles, of tearing away obstructions to get at something desired, of picking valuables out of holes and pockets, of assaulting a neighbor who has something desirable or who has caused pain or who is in the way, of deserting family and other relatives, and of promiscuous sexual relations. (Davenport, 1911)

The term was broad enough to encompass the most disparate categories of "dangerous people."

Among the most important American eugenics publications on feeblemindedness are successive volumes by Henry Herbert Goddard: *The Kallikak Family* (1912) and *Feeble-Mindedness: Its Causes and Consequences* (1914). The former in particular exemplifies the methodological laxness of many eugenics publications. Yet this "genuine story of real people" (presented under a fictitious name) was also hugely influential in public discourse.

Introduced as "a natural experiment of remarkable value to the sociologist and the student of heredity," the story of the Kallikaks is a highly moralistic drama. It begins when an honorable member of a middle-class family "in an unguarded moment steps aside from the paths of rectitude and with the help of a feeble-minded girl, starts a line of mental defectives that is truly appalling" (1913: 50). Degenerate as this line is, "no amount of education or good environment" can change their fate "any more than it can change a red-haired stock into a black-haired stock." In this "primal myth of the eugenics movement" (Gould, 1981: 198), the author concluded:

> No amount of work in the slums or removing the slums from our cities will ever be successful until we take care of those who make the slums what they are.... If all of the slum districts of our cities were removed to-morrow and model tenements built in their places, we would still have slums in a week's time, because we have these mentally defective people who can never be taught to live otherwise than as they have been living. Not until we take care of this class and see to it that their lives are guided by intelligent people, shall we remove these sores from our social life.
>
> There are Kallikak families all about us. They are multiplying at twice the rate of the general population, and not until we recognize this fact, and work on this basis, will we begin to solve these social problems. (1913: 70–71)

Amelioration, utopia, and plasticity: revising the eugenic ethos

From the perspective of Goddard and many others, eugenics looks very much like a form of biological classism and racism, obsessed with gloomy fantasies about otherness and its dangers. However, this dark imagined landscape of fear and social resentment was counterbalanced by a more activist discourse centered on visions of human advancement.

Unlike evolutionary psychology and other forms of later biological determinism based on the presumed fixity or permanency of the human condition, eugenics could be futuristic, even utopian. From the start, a substantial group intellectuals and scientists professed this future-oriented eugenics in tension with more conservative strands of the movement. As Alison Bashford and Philippa Levine write:

> If eugenics was about the problems of inheriting the past, it was also about the optimistic possibilities of planning future generations. There was a power in eugenic promise – perfectibility, improvement, the benefits that would accrue from rational planning. Despite the persistence of a degenerationist discourse, eugenics was thus marked by considerable optimism.... Meliorist terms such as "race betterment" and "race improvement" were titles commonly chosen by and for eugenic associations, especially those with a greater lay and community membership. Eugenics was premised on a belief that science was of necessity reformist in its intentions and aspirations. (Bashford and Levine, 2010)

Indeed, the ameliorist rhetoric was widespread; left eugenicists did not have exclusive purchase on it. The study of heredity, said Davenport, "stands as the one great hope of the human race, its savior from imbecility, poverty, disease, immorality" (1911). Galton himself emphasized progress, not degeneration. "The first pedigrees Galton composed were not of epileptic families," Bashford and Levine point out, but of talented ones, "the Wedgwood-Darwin-Galton family to which he himself belonged; these studies traced the inheritance of ability" (2010). Many eugenicists were committed to the plasticity and perfectibility of human nature, a natural ideological complement to the idea that the future of evolution could be taken into human hands and made to comply with human wishes.

By controlling evolution, humanity would achieve its potential. Focusing on these characteristics – future-oriented attitude, social planning, and capacity of human nature for social manipulation – makes it possible not only to understand eugenics in a less monolithic sense. It also helps, more importantly, to understand eugenics as a specific and in a sense unique ethos. Although eugenics as a *discipline* survived its supposed widespread discrediting by Nazism and persisted in complex, often subtle ways after 1945, penetrating many debates (abortion, overpopulation, genetic counseling, prenatal testing and preimplantation genetic diagnosis etc.; see Bashford, 2010) I claim that eugenics' social engineering, utopian and quasi-religious ethos practically died around 1940 (with a residual exception in the rhetoric of the transhumanist movement), so that it becomes possible to clearly separate two different thought-styles, in the political use of biology before and after World War II.

Moreover, focusing on the specific social engineering ethos of eugenics in the first four decades of the last century helps distinguishing it from the other forms of biologism that preceded it, like the individualistic and laissez-faire based "social Darwinism." Eugenics as an ethos of political radicalism with a social engineering vein and the insolence of scientists as planners of a new society, as experienced internationally in the period 1900–1940 therefore represents *a historical unicuum.*

A secular religion: eugenics as a utopian ethos

In the context of the broader revision of the history of eugenics summarized above, over the last few decades there has been an increasing effort to highlight the utopian dimension of the eugenic experience. The link between the construction of a utopian society and the social control of human breeding goes back, after all, to Plato, Campanella, and More (Hertzler, 1923; Bloomfield, 1949). This link has been recognized as an integral, not merely superficial, part of the writings also of modern eugenicists from Galton to Huxley, from Muller to Haldane and Fisher (Morton, 1984; Parrinder, 1997; Adams, 2000; Esposito, 2011).

Eugenics, according to Peter Morton, was "a component in most of the Utopian writing after 1870" (Morton, 1984). As Diane Paul has noted, literary works, such as George Bernard Shaw's *Man and Superman* (1903) and H.G. Wells's *A Modern Utopia* (1905) contained explicit eugenic themes and "probably did more than any academic studies to popularize the concept of selective breeding" (Paul, 1995: 75; see also Hale, 2009).

In an 1865 lecture, Galton referred for the first time to a Utopia

> in which a system of competitive examination for girls, as well as for youths, had been so developed as to embrace every important quality of mind and body, and where a considerable sum was yearly allotted to the endowment of such marriages as promised to yield children who would grew into eminent servants of the State. (quoted in Pearson, 1930/2011: 78)

Addressing the 1891 International Congress of Hygiene and Demography, Galton presented eugenics as an effort to raise the "present miserably low standard of the human race" to one in which "the Utopias in the dreamland of philanthropists may become practical possibilities" (Pearson, 1930, 3A: 220):

> I wish again to emphasize the fact that the improvement of the natural gifts of future generations of the human race is largely, though indirectly, under our control. We may not be able to originate, but we can guide.... It is earnestly to be hoped that inquiries will be increasingly

directed into historical facts, with the view of estimating the possible effects of reasonable political action in the future.

And in 1901, at his Huxley Lecture on The Possible Improvement of the Human Breed, Under the Existing Conditions of Law and Sentiment, Galton "wryly confessed that he had 'indulged in many' utopias, though the one that he announced – his dream of a missionary organization keeping statistical registers of the nation's heredity – can scarcely have captured his listeners' imagination" (Galton, 1909: 33).

Galton fantasized about such schemes for his entire career, unto the end of his life. Their fullest realization is in his unpublished utopian novel *Kantsaywhere*, which he worked on in 1910, at the age of eighty-eight. Although only a fragment, and of poor literary quality (see Morton, 1984), the novel revealed one of his profound ambitions: not just to correct the irrationality of reproduction in English society but also to imagine a new polis in which the mechanisms of human heredity were entirely under the control of eugenically oriented institutions. In the land of Kantsaywhere, reproductive functions are "regulated by an oligarchy selected by tests" with power placed in the hands of "a eugenic corps d'elite or caste" (Blacker, 1952b). Immigrants are medically checked, the unfit barred from reproduction, and the "very inferior" sent to labor colonies. A state council, the Eugenics College, uses "a system of universal education and certification" to test the qualities of citizens and intended spouses (Pearson, 1930, 3A: 416).

Along with these utopian/dystopian discourses, the first generation of eugenics writings incorporated a messianic message of replacing "the old man with a new Apollo" (quoted in Pearson, 1930). Thus Galton compared eugenics, a "virile creed," to "'Jehad,' or Holy War against customs and prejudices that impair the physical and moral qualities of our race" (1909). He even wished that eugenics would become "a religious dogma among mankind," that it would be "introduced into the national conscience, like a new religion" (Galton, 1909, 68; see also Saleeby, 1906).

Issues of *Eugenics Review* were filled with references to eugenics as a modern form of religion, compatible with Christian faith as both were based on the goal of reaching human perfection (Inge, 1909). These religious and even soteriological overtones were clear to critics of the eugenics movement, from the Darwinian Wallace to the Catholic Chesterton. They sensed in the eugenics movement the "the meddlesome interference of arrogant scientific priestcraft"

(Wallace in Rockell, 1912), a tyranny of doctors and "tenth rate professors" (Chesterton, 1922).

Bio-utopias after Galton

The utopian coloration of eugenics was even more marked in the generation after Galton, especially among leftists such as Haldane and Julian Huxley, but also with different nuances in Fisher's work.

Haldane's writings are an obvious source of utopian themes, and not only in his most famous futuristic book, *Daedalus: Or Science and the Future* (1924), in which love and sex are fully separated and children are conceived through ectogenesis – outside of the maternal womb. In that work, ectogenesis saves civilization from collapse "owing to the greater fertility of the less desirable members of the population in almost all countries" (from 1923 version of the text read before the Heretics society at Cambridge). As Mark Adams has noticed (2000) utopianism is a key part in a minor 1927 essay, "The Last Judgment," as well. Whereas *Daedalus* anticipates the next hundred and fifty years of human evolution, and

> details how the decline of humanity was forestalled by the universal application of eugenic ectogenesis... the futurological account in "The Last Judgment" is from 40,000,000 years into our future, and details how we have survived the end of our planet and risen to a higher form of consciousness through controlled human evolution. (Adams, 2000)

The key point for Haldane was to replace traditional religion with the new religion of biological science. Again in Adams' words (2000):

> Humanity is at a crucial moment in its history as a species, with only a few centuries remaining for us to seize control of our destiny: left to natural law, humanity will degenerate and, like all other biological species, eventually became extinct; our planet (and later our sun) will die. But the new biology affords us a way out: In the short term, we can halt our degeneration through some form of negative eugenics, social experimentation, world government and technocratic socialism. In the long term, using positive eugenics and bioengineering, we can create new kinds of humans for moving into space and colonizing other planets within – and, if possible, beyond – our solar system. In this way, human progress can proceed indefinitely, producing future descendants with even higher (perhaps telepathic or communal)

> forms of mentality. This is the science-based faith that will provide what Christianity and other religions cannot: scientific answers to the profound questions of ethics, human destiny, our place in the universe and the meaning of life. To realize our true destiny, we must be guided not by a myth from our past but by a vision of our future.

In subsequent texts, including in his 1932 *The Inequality of Man and Other Essays*, Haldane returns to the literally unlimited possibilities derived from taking control of human evolution. In a millenarian style, he claims that once control of human evolution is guaranteed, man will know

> no bounds at all to his progress. Less than a million years hence the average man or woman will realize all the possibilities that human life has so far shown. He or she will never know a minute's illness. He will be able to think like Newton, to write like Racine, to paint like Fra Angelico [or van Eycks in a different version, my note], to compose like Bach. (1932: 144)

Another approach to "millenarian biotechnological dreams" (Esposito, 2011) comes via Huxley's "technocratic utopianism." A key theme in Huxley's worldview is biological progress, often inflected with semi-purposive and certainly anthropocentric themes (Greene, 1990; Smocovitis, 2009). Huxley highlights

> the tendency, as embodied in the facts of evolution, of living matter to progress to ever higher levels of achievement, into forms which have more internal harmony, more external control, more intensity of mental life. And man, with his scale of values, is the culmination of this second trend (Huxley, 1931; see Sluga, 2010).

This view of biological progress is in line with eugenics (Beatty, 1992) because it must be guided by human beings. Moreover, looking to a better future speaks to the present inadequacy and perfectibility of human beings, who therefore require urgent eugenic intervention. These themes pervade *What Dare I Think* (1931) and *If I Were Dictator* (1934), and they adapt themselves to new times after the Second World War (Weindling, 2012).

It should not be forgotten that even Huxley's 1942 classic *Evolution, the Modern Synthesis* ends with a chapter dedicated to evolutionary progress, defined, in technocratic style, as "increased control over and independence of the environment" (1942: 564). His continuing passion for social engineering, for humankind's transformation into an altruistic and cooperative species owing only to a radical change in "reproductive habits," is striking (ibid.: 573, on the "gospel of evolutionary progress" in Huxley, see Greene, 1990; see also; Smocovitis, 2009). But Huxley's faith in progress as a fact of evolution was limited to a "few selected stocks". Therefore improvement was "not inevitable" – implying once again humanity's need of a eugenic strategy to guide its destiny. (It was, ironically, Julian Huxley's brother Aldous who seemed most fully to grasp the moral problems of science as redemptive religion, in his 1932 magnum opus *Brave New World*.)

An even more explicit political inflection of the bio-utopia can be found in the work of the leftist H. J. Muller. In *Out of the Night* (1935), Muller writes of organisms "far more plastic in their hereditary basis than had been believed" and therefore malleable according to the range of human desires (44). As in Haldane's *Daedalus*, in Muller's fantasia, sex and reproduction are radically disjoined. New generations are born thanks to bold techniques such as the transplantation of "the fertilized egg from one female to another" or of a "portion of ovaries from one person to another." It is possible even to grow "small amounts of ovarian tissues outside the body" (108–110). With these methods employed for the selective breeding of the best stocks, "It would be possible for the majority of the population to become of the innate quality of such men as Lenin, Newton, Leonardo, Pasteur, Beethoven, Omar Khayam, Pushkin, Sun Yat Sen, Marx" (ibid.)

The malleability of human biology: human nature as plastic as clay

Since about 1970, sociobiology and then especially evolutionary psychology have pressed the claim of biological determinism, which holds that human nature is, if not immutable, fixed since the Pleistocene. One can retrospectively attribute a similar attitude to eugenicists, but this would be wrong. Not just Muller but virtually all eugenicists thought human nature was plastic. From the beginning with Galton, plasticity was inextricable from eugenics.

After all, if eugenics aims to take control of human evolution, there must be a space for human actions to shape deliberately the biological future of the species. The plasticity of the eugenicist is the plasticity of the breeders, who dream of molding their "products" according to their needs. Opening *Hereditary Talent and Character*, Galton writes

> The power of man over animal life in producing whatever varieties of form he pleases, is enormously great. It would seem as though the

> physical structure of future generations was almost as plastic as clay, under the control of the breeder's will. It is my desire to show more pointedly than – so far as I am aware – has been attempted before, that mental qualities are equally under control.

Control – i.e., eugenics – necessitates plasticity. The notion of the malleability of human evolution, and therefore of human nature, is also implicit in Galton's view that

> the processes of evolution are in constant and spontaneous activity, some towards the bad, some towards the good. Our part is to watch for opportunities to intervene by checking the former and giving free play to the latter. (1869)

In the generation after Galton, soft-hereditarian eugenicists such as Kammerer, whom I will discuss in Chapter 4, and hard hereditarians such as Haldane and Muller also affirmed the plasticity of human nature. Haldane and Huxley wrote in 1927 about this program to take charge of and guide the march of human evolution:

> The one great difference between man and all other animals is that for them evolution must always be a blind force, of which they are quite unconscious; whereas man has, in some measure at least, the possibility of consciously controlling his evolution according to his wishes. But that is where history, social science, and eugenics begin, and where zoology must leave off (quoted in Adams, 2000).

Muller's discovery in 1927 of the mutagenic effects of the x-ray further bolstered confidence in human control over hereditary material and therefore the evolutionary process.

Fisher, the right-wing eugenicist, came even closer to what we would today call "transhumanism," a term coined by Julian Huxley after World War II, and one of the few visible legacies of eugenic pre-WWII spirit of utopia.[3] Fisher, who awaited eugenics as a new religion for the European soil, argued forcefully that human stock could be improved and human nature made a target of intervention and manipulation. In "Some Hopes of a Eugenist" Fisher references Nietzsche's übermensch (see also Searle, 1976; Stone, 2002) and writes:

> We can set no limit to human potentialities; all that is best in man can be bettered; it is not a question merely of producing a highly

> efficient industrial machine, or a paragon of the negative virtues, but of quickening all the distinctively human features, all that is best in men, all the different qualities, some obvious, some infinitely subtle, which we recognise as humanly excellent.
>
> But Darwinism is not content to reveal the possible, perhaps the necessary, destiny of our race; in this case the method is as clear as the ideal; the best are to become better by survival. It is in this that we differ from less biological Utopia seekers; humanity has never been poor in desires, in hopes or in dreams, it is the ways and means, the concrete results that are so sorely inadequate; eugenics comes at an appropriate time, when our civilisation is already sadly acknowledging that the great bar to progress lies in human imperfection; for the first time it is made possible that humanity itself may improve as rapidly as its environment. (1914: 310)

Far from being a static, fixed reality, the figure of the human conceptualized by eugenicists was very much a figure in transition from a present imperfect state, to a better one, and it was the responsibility of science and scientific experts to achieve it with the right measures.

Conclusion: utopian ethos and the first era of political biology

That the eugenic ethos was animated not only by a meliorist, future-oriented vision but even by a utopian philosophy can partly counterbalance the still prevailing association of eugenics with mere political reaction. Reactionary and degenerationist views were obviously part of the repertoire but alongside this upward view: man, and nations, could fall and could rise.

However, the fact that eugenics incorporated a future-oriented vision does not exculpate any of its promoters, especially given post-1945 moral standards. From Galton's consignment of the unfit to labor colonies, to Muller's bold reproductive proposals, the eugenic utopia was entirely dystopian. All these views of the future were based on a totalitarian idea of a fully regulated society "where liberal and democratic principles were partially or totally suspended in favor of bioscientific control and planning for the future" (Esposito, 2011).

It is no coincidence that only recently have scholars brought to light this wildly utopian dimension of eugenics, a dimension that was long neglected, or hidden, after 1945. With the revelation of the Nazi horrors, the internal historiography of the movement erased its utopianism.

Biology had to be compatible with broader values of human rights, universalism, and liberal democracy (see Chapter 5). In this context, there was no place for the utopianism of past eugenic writings (Osborn, 1940; Blacker, 1952b; Maynard Smith, 1988).

For instance, in the name of novel post-1945 prudence, the psychiatrist Carlos Blacker defended Galton's sinister utopia as "amiably paternalistic," naïve rather than authoritarian and not to be taken seriously (1952). Galton was, in sum, joking; his coercive eugenic state had nothing to do with Nazism. In 1972, John Maynard Smith attempted to downplay eugenic utopianism by claiming, "Recommendations of positive eugenic measures can at present only distract attention from more urgent and important questions.... Eugenics can wait; birth control cannot" (1972: 78).

The utopianism of eugenics as it operated globally from 1900 to WWII indicates that the movement was defined less by a theory of heredity and more by a moral ideal. It was an ethos supporting managerial oversight and manipulation of the human-biological dimension, and it crossed scientific and political divisions – between hard and soft-hereditarians and right- and left-wing authors. What united these disparate efforts was a certain view of the relationship between scientific expertise and society. Eugenics was a view of "scientific activism.... in the hope of improving humanity" (Adams, 1990b: 219).

In all its political and epistemic manifestations, eugenic scientific activism was imbued with the values of aggressive social interventionism. It was an authoritarian high-modernist project in the sense articulated by James Scott, an ideology of rationalization of society and nature based on a "muscle-bound version of the self-confidence about scientific and technical progress" (1998). Science and expert knowledge were unproblematic, indeed supreme values. Another way to look at eugenics in this pre-1945 phase is through Zygmunt Bauman's category of the "garden-type" politics, or "gardening state," (Mottier, 2010; see Bauman, 1989; Mottier, 2008). For Bauman this exquisitely modern posture implied an understanding of society as a "garden to be designed and kept in the planned shape by force", "an object of designing, cultivating and weed-poisoning" (Bauman, 1989). Indeed, weeding and cultivating the human condition, and getting rid of its bad seeds, was eugenics' profound ethos – selecting the fit and sterilizing the unfit or, in a Lamarckian version, secluding the degenerated and regenerating the weak. For instance, the philosopher and eugenicist Ferdinand Schiller explicitly advocated such a gardening posture when he urged to taking care of the "human weeds" so that humanity might "cultivate our garden" (Schiller, 1930). Such metaphors were widespread in eugenic publications, which promoted human gardening vigilance as part of a broader idea, common to eugenics and racial hygiene, in which biology served as a source of social prophecy and regeneration (Weindling, 1989).

Eugenics, racial hygiene, and the larger modernistic culture of social planning and social engineering demarcate the political biology of 1900–1945. Before 1900, the relationship between biology and politics was mostly social Darwinism (which largely coincided with the Gilded Age in US History, 1870–1900).[4] Often confused with eugenics, or wrongly understood as an anticipation of eugenics, social Darwinism, a (contested) label originating in late 1870s (Bannister, 1988) and made popular by Hofstadter's work (1944; see for a critique: Bannister, 1988; Crook, 1994; Paul, 2006; Leonard, 2009), in fact diverged from eugenics in many ways. Two of these are especially significant in the current context.

First, social Darwinism was contemptuous of government action. It supported a minimal state and a laissez faire culture of competition among individuals (as an extrapolation of intra-species struggle for survival). This sort of politics directly contradicts that of the eugenics movement, which advocated massive, state-directed intervention in order to bolster the health of the race. Where social Darwinism preferred individualism, eugenics subordinated the individual to the broader goal of racial improvement. Thus there could be a leftist, socialist eugenics but not a leftist, socialist "social Darwinism".

Second, at the core of social Darwinism was a Lamarckian belief in "the benefits of hard-won experience." Continuous progress was seen to result from improvement of the species generation by generation (Bowler, 1989: 11). According to Walter Bagehot, banker, journalist, and a key figure of British social Darwinism, history is "a science to teach the law of tendencies – created by the mind, and transmitted by the body – which act upon and incline the will of man from age to age" (quoted in Paul, 2006). This was a classical Lamarckian and in its own way progressive philosophy of history, based on habits turned into instincts, something nearly completely missing in the eugenic conceptual landscape. This view only lingered in the background of soft-hereditarian eugenics that, moreover, tended to emphasize degenerative rather than progressive processes (One exception was Kammerer; however, his ethos was not individualist and competitive, as in social Darwinism, but rather cooperative and socialist.). Whereas social Darwinists thought the habits of successful people could be turned into the biology of the next generation, mainstream eugenics thought there was no way to improve the human race by "stimulating individuals to greater effort" (Bowler, 1989: 11). Only by changing the balance between good and bad genes could society be improved. As Julian Huxley claimed in his 1936 Galton Lecture, eugenics emerged as a reaction to "the intellectual excesses of the perfectionists and sentimental environmentalists, who adhered to the crudest form of Lamarckism and believed that improvements in education and social conditions would be incorporated in an easy automatic way into human nature itself and so lead to continuous and unlimited evolutionary progress" (1936).

The differences between eugenics and social Darwinism are further illustrated by their attitudes regarding war. Social Darwinists, largely imperialist in their approach to international politics, believed in war as an accelerator of the struggle for existence. For eugenicists, war was mostly a huge waste of the nation's best germ-plasm (Adams, 2009).

When we understand eugenics as a modernist ethos of social engineering and social planning, subordinating the individual to the good of the race, and moving in a gardening attitude in the sense of Bauman (1989), we also see how the first period of political biology, 1900–1945, differs from what came after World War II. Eugenics persisted after 1945, for instance in debates over birth and population control (not to mention, obviously, in the persistence of coerced sterilization until 1970s in several North American and European countries). But in its Western post-war manifestation, the wild utopianism of pre-war eugenics was tamed, constrained by the checks of liberal democracies and by the universalism of human rights.

This does not mean that the post-1945 liberal-democratic framing of biology was bereft of exclusionary or racist potential. But none of the four defining elements of the eugenic ethos – radical biologism, utopian social engineering, unlimited empowerment of scientific experts, and primacy of race over individual – lay at the heart of the post-war political biology, as I will show in Chapters 5 and 6. Said in passing, the problem of post-1945 eugenics was exactly an identity issue: whether anything like eugenics could still exist in a society that got rid of these four axes (that is whether a eugenics based on individual choices, suggested but not directed by experts, and limited in its scope to a few physical traits, without any plan to change society, could still be a eugenics).

However, before moving to the great changes wrought by World War II, we must zoom in on interwar eugenics to see more precisely the richness and polysemy of debate on human heredity and its potential for a cartography of our present.

4
A Political Quadrant

Lost political-epistemic options before "crystallization of values"

Until 1930 circa, the debate around the sociopolitical implications of human heredity was complex and pluralistic. Though today eugenics is often thought of, especially in social-science quarters, as an exclusively hard-heredity right-wing phenomenon, at that time, a vast array of hereditarian political-epistemological philosophies aimed at control of human evolution.

Over time, this variety was lost. In theories of human heredity the early twentieth century was, in a sense, a classical moment of conceptual turbulence, and therefore extreme richness and differentiation of positions, *before the controversy was closed*. However, in this case, the controversy was not merely scientific – the soft-versus-hard-heredity debate – but more broadly political-epistemological. Both right and left could politicize soft and hard heredity.

In scientific discussions – especially among biologists, as opposed to doctors and educators – hard heredity was becoming the mainstream position, but it was still challenged within eugenics circles. On the political side, nature and nurture were not yet unilaterally aligned and logically associated with conservative and progressive political values, respectively. In Europe and North America, this scenario would only solidify after about 1930 (see for instance Pastore, 1949). However, before such a solidification, things were much more plural. Under a soft-heredity framework, racist and reactionary political values could be mobilized on behalf of nurture and the environment. Conversely, egalitarian, radical, and even overtly communist discourses were constructed under the banner of strictly hard-hereditarian eugenics.

Opinions on the social consequences of heredity can be plotted as an intersecting quadrant of politics and science: right and left Mendelism (or, better, hard-hereditarians)[1] and right and left neo-Lamarckism (soft hereditarians) (see Figure 4.1). In all these cases, scientific views mattered to politics, and politics was shaped by scientific views. Left Lamarckians were not left Mendelians, nor was the understanding of degeneration for right-Lamarckians the same as for right Mendelians.

If this political-epistemological quadrant seems novel, it is only because we have forgotten the pathways by which heredity was politicized in the past. After the controversy was closed, a specific alignment of scientific and political values – right Mendelian – crystallized, to draw on terminology developed by historian of science Loren Graham. There was no "pre-ordained logic" in the alliance between eugenics and conservatism, Graham writes (1977: 1158). Similarly, other historians have shown not only that hard-hereditarian eugenics could be associated with leftist values, but also that the emergence of environmentalism as a left-wing and egalitarian ideology was the result of a particular crystallization in the 1930s (Paul, 1984, 1995; Freeden, 1979; Bowler, 1983/2009; Mazumdar, 1992; Staum, 2011), though obviously there is a prequel in the nineteenth century to this story, (see Paul and Day, 2008). Also neo-Lamarckians and authors emphasizing the primacy of nurture could be racist and right-wing.

Crystallization is a particularly useful metaphor. It implies a dynamic process whereby sociopolitical values previously untethered from any

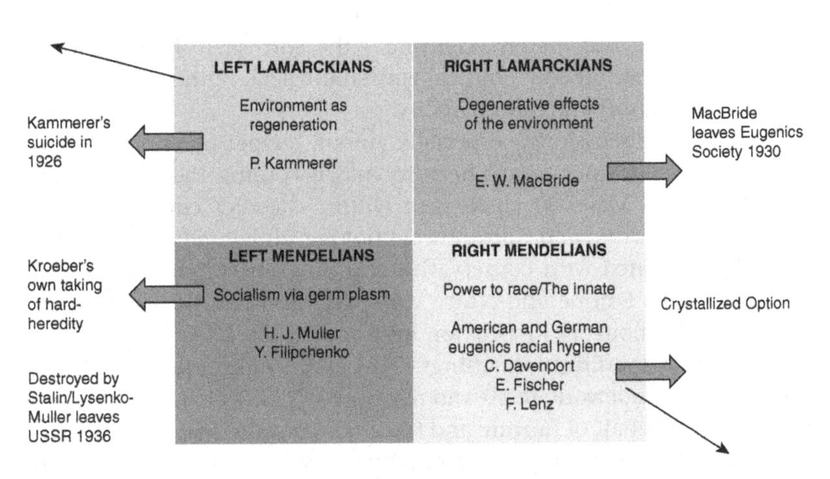

Figure 4.1 Political quadrant of eugenics: ca. 1900–1930

scientific stance are stabilized and form "natural" bonds with certain scientific positions. Upon these apparently natural political-epistemic assemblages, future generations build the arsenals of conceptual warmaking. But the seeming logic of the alliance is only the *a posteriori* effect of a contingent stabilization: things could have been different. Thus it is critical that historians excavate (or "problematize," Foucault, 1984, 1985) the turbulent moment before crystallization. I aim to show the multiplicity behind the politicization of human heredity because some of the early twentieth century's possible alignments – those of right and left Lamarckians – can return to us today, filtered through the lens of contemporary biology, in particular epigenetics with its claims of a return to soft heredity.

My analysis of the political quadrant begins with the soft-hereditarian axis, in its two versions, right- and left-wing. This axis existed until about 1930, after which it petered out, leaving no remaining European soft-hereditarian eugenicists (amongst the professional community of biologists; things are different for doctors and educators, puericulturists and social reformers, and are very different out of Europe and North America). In Europe, leftist soft heredity was closed by Paul Kammerer's suicide in 1926 (Gliboff speaks of a Golden Age of Lamarckism ending in 1926, see 2011); Ernest MacBride's resignation from the Eugenics Society in 1931 was the end of a coherent right-wing Neo-Lamarckism.

Soft heredity would survive in an organized form in biological circles mostly within the Lysenkoist ranks in the Soviet Union, though there were isolated exceptions in the West. But Lysenkoism, for ideological reasons, explicitly vetoed any applications of its biological science to human heredity. There was in the post-1930s Soviet Union an institutionalized soft-hereditarian biology, but it was not eugenic.

After exploring the two politics of Lamarckism, I move to the second axis of the quadrant, where I focus on the left-Mendelian position, looking in particular at Muller and the short-lived experience of Soviet eugenics. Instead, I am not concerned here with the right-Mendelian quadrant, which mostly coincide with what Kevles called as "main-line eugenics" (1985). In the previous chapter, I covered elements of right-wing Mendelism (and broadly hard-hereditarianism) – the classist and racist agendas of Anglo-American eugenics – and I will discuss the German racial hygiene movement before and under Nazism in the next chapter. But, more importantly, I have less methodological interest in right Mendelians because theirs was the crystallized option in the international movement (Kühl, 2013); that history does not require excavation.

The three rival tracks were lost, and eugenics, nature and in a sense the same idea of a public use of genetics ended up aligning with right-wing values. This alignment pushed nurture, soft heredity, and the rhetoric of the environment to the left of the political spectrum. This was obviously a possibility before that crystallizing moment (see Paul and Day, 2008), but not a pre-ordained one. This supposedly natural but actually constructed (and contingent) politicization of scientific concepts formed the apparently logical background of debates on the boundaries of biology and society and biology and politics for the all twentieth century, long after the dust of this debate had settled and its complexity and pluralism forgotten.[2]

Right-wing neo-Lamarckians: pathogenic environment and the rhetoric of degeneration

Lamarckian ideas are often associated with social reform, if not true political radicalism and socialism (Desmond, 1989; see also Koestler, 1971). As historian Gianna Pomata writes:

> Soft, environmental hereditarianism was inherently optimistic. Precisely because the original cause of hereditary disease was environmental, its transmission could be stopped by behavioural and environmental changes brought about by preventive medical measures. Hereditary diseases could be "disinherited" thanks to behaviour properly regulated by medical advice. (2003)

In a sense, there appears to be a potential linkage between Lamarckism and egalitarian, left-wing, and future-oriented views (see also the important work of Gissis, 2002)[3]. Engels for instance was a believer in the inheritance of acquired characters (1896)[4], and in 1910 the association between Marxism and soft heredity was perceived so evident that S. Herbert protested in the *Eugenics Review* that Marx's "followers have thoughtlessly extended the meaning of this axiom by asserting that it is the 'environmental' factor which is all-important, determining, as it does, man, his morals, character, and even his progeny. They have become Lamarckians, believers in the inheritance of acquired characters, on the grounds of their political convictions. According to their idea, it is only necessary to alter our present economic structure in order to effect a radical change in the physical, mental and moral condition of the people" (1910).

Stalinist Soviet Union – mostly for ideological reasons – made their version of Lamarckism, the Michurinian-Lysenkoist doctrine, the official party line in biology until the mid-1960s. As V. V. Babkov reports in his important volume on the history of Soviet eugenics, Lamarckism and the left became so intertwined in the Soviet Union that the proceedings of a 1930 conference between geneticists and Lamarckians misprinted "Lamarckian schools" as "Lamarxist schools" (2013: 534).

Lamarckism's "natural" affinity with the left seems to emerge even more looking *a contrario* at how Nazis embraced anti-Lamarckist Weismannism. In Nazi Germany, Lamarckism and the inheritance of acquired characters were considered a malicious trick played by Bolsheviks, liberals, and of course Jews, all seeking to undermine the fundamental power of race by attributing primacy to the environment. Nazi geneticist Fritz Lenz described the Jewish embrace of Lamarckism as an obvious "expression of the wish that there should be no unbridgeable racial distinctions." He sarcastically added, "Jews do not transform themselves into Germans by writing books on Goethe" (quoted in Proctor, 1988: 55).

So it may therefore come as a surprise that conservative and racist neo-Lamarckians had considerable influence. As scholars including George Stocking (1968) and Peter Bowler (2009) have argued (see also Haller, 1971), in spite of its "untarnished image as the reformers' biology," neo-Lamarckism was a major element of a racist and classist nineteenth-century agenda that continued through the early part of the twentieth (Bowler, 1989: 156; see Bowler, 1983). In fairness to Lamarck, however, we need to add that their strand of Lamarckism was a truncated version of the inheritance of acquired characters, which emphasized the passive reception and transmission of deleterious features rather than the acquisition of positive ones in active response to their environments.[5]

There are at least three reasons why social scientists have ignored conservative neo-Lamarckism. First, it is difficult to define Lamarckism, given the vast circulation of soft-hereditarian ideas in the nineteenth century – that confusion of nature and nurture so ably described by historians such as Carlos López-Beltrán (2004). Loosely Lamarckian or soft-hereditarian ideas were everywhere in medical and anthropological debates, so widespread that there was almost no need to formulate them carefully (Stocking, 1968). Second, often authors who did explicitly formulate racist Lamarckian views – such as the French anthropologist Paul Topinard, a follower of Broca in the craniology tradition (Staum, 2011) – held otherwise progressive positions on social matters (opposing for instance the extreme atavism of Lombrosians; see Nye, 1984; or believing in the possibility of inculcating altruism, Staum, 2011). Thus it can be hard to isolate conservative Lamarckists from progressive ones. Third, neo-Lamarckists were ideologically fragmented (Burkhardt, 1980) – and, lacking the unifying core doctrine that held, say, Mendelians or biometricians together.

To overcome these interpretive problems, we must define clearly right-wing neo-Lamarckism. One essential component of this line of thought is the configuration of environment as *a constant source of morbidity* that may permanently *alter, weaken,* or *poison* heredity. But this scientific stance alone does not describe a political agenda, because it could generate a reformist agenda to reverse these pathogenic environmental effects.

Thus we must look to the author's aims. Was the goal to reverse negative effects? Rarely. More often, this constellation of right-wing Lamarckists viewed morbid environments as uncorrectable sources of hereditarian disruption. They therefore leaned toward anti-egalitarian and exclusionary discourses in which social and racial groups too long exposed to toxic environments were seen as a biologically damaged underclass or inferior race whose offspring were condemned before birth. In Europe and the United States, this discourse was often the basis of calls for citizenship restrictions.

The definition, then: to be classified as right-wing neo-Lamarckist (or right-wing soft-hereditarian), an author must *emphasize the pathogenic qualities of environmental influences and the acquisition of harmful characteristics and must believe that such deleterious processes leave affected groups socially irredeemable, or less fully citizens.*

The first school of thought to fully comply with these criteria is medical degenerationism, which was widespread in the medical and social literature of the nineteenth and early twentieth century. Degenerationists believed that a pathogenic environment acts as a "racial poison" (in a late terminology). There were many such poisons: alcohol; sexual diseases; the industrialized, filthy, and overpopulated metropolis, particularly its slums full of ignorance and vice. Doctors, educators, and social reformers were obsessed with the transgenerational perpetuation of toxic environments and bad habits within poor families and dangerous groups. They were convinced that these germ-weakening effects could become fixed degenerate instincts in the new generations where they emerged as forms of nervousness, insanity, and idiocy. Alcoholism was the central foe of this degenerationist discourse (Bynum, 1984). Many doctors saw environment as a first cause of disease and heredity as an accelerator (Coffin, 2003; Staum, 2011), but, on the whole, degenerationists explicitly cited the transmission of acquired characteristics (Bynum, 1984). As Ian Dowbiggin writes, "Degeneracy theory was the medical counterpart to Lamarckian biology" (1991: 148), or in Snait Gissis' term, the notion of degeneration was "subsidiary" to the idea of direct adaptation to the environment (2002).

When Weismannism and hard heredity became prominent after the 1880s, polarizing the heredity debate, what we would consider today soft hereditarians increasingly incorporated degenerationist tropes. Examples such as chronic alcoholism were used to show how Weismann's sequestration of the germ plasm was in fact full of possible holes, exposed its risks of racial poisoning and permanent damage for future generations (it has to be noted in passing that Weismann himself didn't deny the possibility of a direct teratogenic effect of alcohol but claimed that such cases "have nothing to do with heredity, but are concerned with an affection of the germ by means of an external influence" (quoted in Bynum, 1984)). Needless to say, it was the germ plasm of certain classes and races that were most at risk of such poisoning.

A distinctive Lamarckian version of racism followed medical degenerationism. Lamarckian racism thrived on the idea that the inheritance of acquired characteristics and direct environmental influences were the pivotal mechanisms in the formation of races. Races were shaped by the transmission to offspring of locally determined adaptations (Stocking, 1968; Haller, 1971; Bowler, 2009). Following degenerationist views, inferior races were seen as the invariably passive recipients of particularly morbid or unfavorable environmental effects: "Degeneration by environmental influences could account for differing physical appearance and customs" (Jackson and Weidman, 2004: 29). The languid climate of the tropics, for instance, forged the indolence ("lazy and shiftless," Ellwood quoted in McKee, 1993: 75) of certain African races, which arrested their evolutionary development (Bowler, 1994). These races were always at risk of degenerating "towards a more primitive level of organization" (Bowler, 1995: 115; Bowler, 2009).

In addition to these general arguments about race formation, neo-Lamarckians deployed two unique racist strategies. The first is what Staum (2011) has called "differential Lamarckianism" (see also Haller, 1971; Bowler, 2009; Gissis, 2002), the idea that the positive habits and the benefits of moral progress and education could be inherited by the higher races or advanced cultures but not by "the most 'retarded' or innately less endowed non-Europeans." Differential Lamarckianism was particularly salient in the work of many French Lamarckian anthropologists, for instance Topinard,who argued that the environment operates "differentially, depending on the inherent nature of races." His bottom line was that "Europeans, but not inferior races, could inherit moral habits" (Staum, 2011).

Herbert Spencer (in his 1876 *Comparative Psychology of Man*) made similar arguments when he wrote of the "relative plasticity" of different human races, with the most developed (i.e. European) being the "most plastic":

> Many travelers comment on the unchanging habits of savages. The semi-civilized nations of the East, past and present, were, or are, characterized by a greater rigidity of custom than characterizes the more civilized nations of the West. The histories of the most civilized nations show us that in earlier times the modifiability of ideas and habits was less than its at present. (quoted in Jackson and Weidman, 2004: 81)

As Jackson and Weidman summarize, "Imitation could not overcome the ingrained habits of the race. The behaviors carved into the savage's system by Lamarckian inheritance would overpower his puny attempt to imitate his betters" (2004: 83). This double standard – progress for some but not for everyone – gave rise to the typical mixture of optimism and pessimism in neo-Lamarckian discussion of race and social progress.

Along with this ambivalence comes the second exquisitely neo-Lamarckian way of racism. We need to keep in mind that neo-Lamarckians inflated the concept of race to include not only the blood or germ plasm but also social habits and cultural traditions. Cultural habits became fixed into biological instincts and milieu influenced blood (Stocking, 1968; Bowler, 2009); the boundary between the social and the biological was porous. This enlarged racial terrain was fertile ground for racists.

For instance, social scientists took from Lamarckism an argument against the possibility of racial integration. The weight of past traditions became, in this discourse, an obstacle to present moral progress. Not a defective germ plasm but the embodied persistence of habits of thought made it impossible to overcome racial hierarchies. This racist version of degenerationism resonated among American sociologists under Spencer's influence, such as Charles Ellwood, at the University of Missouri, who wrote:

> The negro child, even when reared in a white family under the most favorable conditions, fails to take on the mental and moral characteristics of the Caucasian race.... His natural instincts, it is true, may

> be modified by training, and perhaps indefinitely, in the course of generations; but the race habit of a thousand of generations or more is not lightly set aside by the voluntary or enforced imitation of visible models, and there is always a strong tendency to reversion. The reappearance of voodooism and fetishism among the negroes of the South, though surrounded by Christian influences, is indeed to be regarded as due not so much to the preservation of some primitive copy of such religious practices brought over from Africa as to the innate tendency of the negro mind to take such attitudes toward nature and the universe as tend to develop such religions[6]. (1901, quoted in Stocking 1968; see on the complexity of Ellwood position: Cravens, 1971; Degler, 1991; McKee, 1993; Breslau, 2007; Turner, 2007)

Through these discourses, neo-Lamarckism contained great potential for racism and classism. Not all neo-Lamarckists were thoroughgoing racists, though. More often than not, their views consisted of a mixture of liberalism and exclusion befitting the ambivalent nature of soft inheritance (see also McKee, 1993). But no matter how much social optimism or pessimism was mixed in, neo-Lamarckism offered a conceptual basis by which to claim that specific groups or races were damaged by their long-term exposure to poor environments or habits. Their germ plasm was permanently poisoned or weakened by the influence of these experiences. The inertia of past legacies was destined to prevail over present reform efforts. This conceptual arsenal would be vital to the soft-hereditarian right-wing of the eugenic movement.[7]

The 1910–1911 alcohol controversy: political ambiguities of nurturism

Caleb Williams Saleeby (1878–1940) was a Scottish doctor, chairman of England's National Birth-Rate Commission, and vice-chairman of the National Council for Public Morals. He was one of the most vocal and eccentric figures in British eugenics, representing an alternative nurture-based line of the movement. As a physician rather than a professional biologist, Saleeby adopted a flexible understanding of heredity that changed across different publications; it is difficult to effectively frame his position (see also Stone, 2002). He recognized the importance of Galton's and Weismann's hard-hereditarian arguments (Rodwell, 1997) but also highlighted several phenomena that in practice escape sequestration of the germ plasm. The two sides of the dispute on heredity (i. e. hard and soft), he believed, were "fooled by the words" (1910a: 37). He was, in sum, a pluralist about heredity, who believed that Weismann and Lamarck could be reconciled (although he is often deemed a Lamarckian, though incorrectly, see Adams, 1990; Stepan, 1991; Woiak, 1998).

This pluralistic view of heredity had implications for his eugenic views. The believer in eugenics, he wrote, needs to dissociate himself from the "cardinal assumption" of neo-Darwinians that "nothing can alter" the germ plasm (1914: 16). The eugenicist has been "too apt to accept without analysis the modern rejection of Lamarckianism, believing that 'acquired characters are not transmitted'" (1910a) thus embracing that "fantastic neo-Darwinian biology which asserts...that parental nurture does not affect offspring." It was therefore necessary to take a less rigid view of heredity in which "some influences affecting future parents will affect the character of their offspring" (1914: 15). These influences, however, were in his case mostly degenerative.

Saleeby's philosophical sources were also peculiar for a British eugenicist. He dedicated his *Progress of Eugenics* (1914) to "teacher and friend" Henri Bergson, illustrating the vitalist, anti-mechanistic influence in his work (Rodwell, 1997). Through his unusual mixture of Galton and Bergson, Saleeby wanted to make of eugenics "more than a glorified materialism." We "want not germ-cells but people," he wrote (1914: 24).

A further feature of Saleeby's work was his distaste for biometry, a science that "measures everything but life" (1910b). Karl Pearson was his archrival and one of the main targets of his invective: Saleeby often described him as a fanatic and an armchair eugenicist who lacked any sense of reality and humanity. Pearson (hard-hereditarian but non Mendelian) represented "nature first" eugenics, whereas Saleeby endorsed what he called a "nurtural eugenics," conceived as "the sum of all the influences which nourish, mould, and modify the individual" and which therefore included education, social reform, and philanthropy. These progressive projects were not, in this schema, antithetical to eugenics (1914: 24, 33).

Saleeby's nurtural view may be mistaken, *prima facie*, for a more liberal form of eugenics, albeit always framed in moralistic language. And he certainly did have more progressive views than mainstream eugenics allowed. He attacked the class bias of many eugenicists and asserted the importance of education and adequate nurturing for every child. He was also active in promoting votes for women, though he insisted on the eugenic role of mothers. And his ideas about segregation and sterilization were undoubtedly more humane than Pearson's.

For his part, Pearson expressed no egalitarian ideals – quite the opposite – and in no way can he be defined a leftist thinker despite his self-proclaimed socialism and his progressive (for the time) views on gender. His positions were among the harshest in the eugenics movement; revealingly, he argued that a high infant mortality rate was a net benefit for the population as a whole. Saleeby mocked his position as the "better dead" school: given the radical imbalance between heredity and environment, for hard-hereditarians (non-Mendelians) à la Pearson the best solution to social problems was to get rid of the defective stocks and not waste time with reforms. Saleeby preferred "preventive eugenics," defined as "the protection of parenthood from the racial poisons," in parallel with preventive medicine (1914: 31; for analogy with Latin American eugenics, see Stepan, 1991).

No matter his good intentions, though, Saleeby's belief that pathogenic agents such as alcohol could poison heredity meant that, in practice, his soft-hereditarianism was close to biosocial degenerationism. The children of alcoholics were not only born already damaged but on account of their acquired intoxication also were "on the average less capable of citizenship" and therefore in need of extended social control (1909). It is true that Saleeby wrote on race regeneration (1911), not just degeneration. But he mainly hoped to prevent racial poisoning.

Like right-wing soft hereditarians, he was obsessed with degeneration and relatively little attracted by the beneficial effects of the environment. As Woiak acutely notices, he "never suggested... that the increased stature or fitness of individuals who had benefited from sanitation or welfare reforms might be transmissible to offspring" (1998: 321). When it came to discussing positive influences, Saleeby denied the possibility of transmission. In summary, his case illustrates how a nurture-first strategy could easily merge with a "fanaticism for eugenics" (Rodwell, 1997).

Both Pearson and Saleeby claimed to be the authentic interpreter of eugenic thought and, especially, of Galton's legacy. This contest for the spirit of eugenics was perfectly reflected in their 1910–1911 debates on alcoholism. Saleeby led the so-called temperance doctors group, which argued that alcohol had degenerative effects on the germ plasm of present and future generations and called for prohibition. Pearson campaigned passionately on the other side. Heredity was hard; there was no such a thing as an alcohol-triggered racial poison capable of producing filial degeneracy.

The debate (covered by Farrall, 1970, chapter 7; Searle, 1979; and especially Woiak, 1998) was sparked by the publication in 1910 of a study by Ethel Elderton, a researcher at the Galton Laboratory, under Pearson's supervision. The results were counterintuitive, provocative, and controversial. The report overtly challenged the evidence and arguments of the temperance doctors and denied the existence of a relationship "between parental alcoholism and defective health in the children." Elderton and Pearson claimed that "the children of the intemperate are healthier than the children of the sober" – an indirect effect, they said, of the stronger constitutions of people who were tempted to drink. They also challenged the common-sense view that defects in offspring – for example, degenerative eyesight – were connected to parental drinking. There was, they said, "a larger percentage of normal eyes among the children of drinking parents than among the children of sober parents" (Elderton and Pearson, 1910).

Uproar followed. Given the vast public policy implications of their argument, the controversy spilled over to newspapers and professional journals. A vast group of doctors, social reformers, and social hygienists rejected Elderton and Pearson's conclusions. John Maynard Keynes took part in the debate, criticizing some of the study's statistical grounds (Farrall, 1970; Woiak, 1998). The study did have its defenders, though, including the demographer Alexander Carr-Saunders (Farrall, 1970). He was among the few who focused on the key point in the alcohol debate: There was no evidence about the inheritance of harmful characteristics, no evidence of filial degeneration. The temperance doctors' claim that "alcoholism on the part of the parents will cause the children to be by nature inferior both physically and mentally" was not persuasive. With an eye to broader debates on hard and soft heredity, Carr-Saunders concluded that the study contradicted "the view that children of alcoholics are inferior as far as inborn characters are concerned" (1911).

In addition to ideological matters, there was also at stake in the debate a struggle for hegemony over "the lines of feasible social reform" within British eugenics (Elderton and Pearson, 1910; see Woiak, 1998). Pearson and his lab represented the side of negative eugenics: If one wanted real knowledge on how to reform society, and not just "untrained philanthropy," one had to mind defective inborn qualities, not environmental factors. Their point was that alcoholism was at most a superficial symptom of inborn defects, certainly not a poison of the human germ plasm.

Saleeby represented the other side of the debate, pushing the idea that parental drinking would irreversibly destroy the germinal tissues of offspring. A few years before the controversy in England, key names in psychiatry such as Auguste Forel (1908; see Bynum, 1984) had even coined a name for these direct pathogenic effects on the germ plasm: "blastophtoria," or the poisoning of germ cells. Alcohol intoxication was the chief example of direct pathogenic action. "The spermatozoa of alcoholics," Forel wrote,

> suffer like the other tissues from the toxic action of alcohol on the protoplasm. The result of this intoxication of the germs may be that the children resulting from that conjugation become idiots, epileptics, dwarfs or feeble-minded.

But not only that. A child of alcoholic parents was an imbecile, and in spite of his own abstinence would preserve

> the tendency to transmit his mental weakness or his epilepsy to his descendants.... In fact, the chromosomes of the spermatozoid...have preserved the pathological derangement produced by the parental alcoholism in their hereditary mneme, and have transmitted it to the store of germinal cells of the feeble minded or the epileptic, who in his turn transmits it to his descendants. (1908: 36–37)

Forel called this phenomenon *fausse hérédité* – fake heredity – because it represented a perversion, or a degeneration, of what was usually handed on from one generation to another (on how Forel actually avoided the trap of claiming explicitly the inheritance of acquired traits, see Bynum, 1984).

Saleeby moved within this broader medical framework of toxins poisoning the germ plasm "of subsequent offspring" (1914: 212). He defined racial poison as "a substance, of whatever nature, which injures the offspring through the parent or parents, and is thus liable to *originate* degeneracy in healthy stocks." Strychnine and boric acid were deadly only to the exposed individual, but a racial poison "whether or not injuring the individual who takes it, is liable to injure the race of which he is the trustee." Strychnine and boric acid were hygienic problems. Racial poisons were eugenic problems (1910b). Saleeby's racial poisons included alcohol, venereal disease, lead, and morphine. Having practiced medicine in the slums of Edinburgh and York, he also thought of these degraded areas as toxic factors, which "directly conduce to alcoholism and sexual immorality, and thus to racial poisoning and destruction" (1914).

Unsurprisingly, Saleeby was not happy with Elderton and Pearson's report. He accused the authors of having cherry-picked figures. He insisted that there were numerous ways in which alcohol could taint otherwise healthy stocks and plenty of studies supporting his case. He claimed "the old belief in the bad quality of children conceived during drunkenness" (quoting Forel, 1910a) had been proven by a turn-of-the-century Swiss study. In response, Pearson eviscerated the study. Saleeby also quoted a "Norwegian study" (Mjoen) that argued "the enormous increase of idiots came and went with the brandy" (1914: 237). And he was not above dirty tricks. For instance, he claimed to have been "informed that a large German brewery is widely using Professor Pearson's conclusions for the purposes of advertisement" (1910c). Pearson, Saleeby contended, encouraged drunkards to breed.

Saleeby's intentions were good: He wanted a eugenics that would welcome "all agencies that make for better nurture." But it is doubtful that, had he been in charge of a social policy program targeting alcoholism, his measures would have been less exclusionary than any Pearson might have devised. He not only embraced the usual eugenic logic that "the feeble-minded, and the alcoholic, and the insane, and those afflicted with venereal disease, must be so guarded and treated in future that they shall not become parents at all" (1914: 77), but he also called for restricting the citizenship rights of both alcoholics and their children (1909). Like most Edwardian reformers, Saleeby never focused on structural problems such as unemployment or inequality. Instead his concern was "personal failings." These were, he thought, "the more easily remediable sources of infant mortality, physical deterioration, and racial decline" (Woiak, 1998).

Furthermore, if one takes seriously Saleeby's conviction that "ailments acquired by parents because of their alcoholism would appear in their offspring" (Farrall, 1970: 270) and that slums lead directly to alcoholism, the logical result is massive intoxication of the germ plasm of people living in degraded areas. This could only mean that cities were putting out generation after generation of degraded individuals. On Saleeby's accounting, the numbers of present and future defectives would have exploded beyond the worst-case scenario of even the harshest hard-hereditarian à la Pearson.

Meanwhile Pearson's position, which was no less fanatical than Saleeby's, had an unexpected egalitarian potential. As much as he opposed social reform, he also was an enemy of moralistic conclusions and sketchy speculations based on nonexistent evidence. While his unyielding hard-hereditarian discourse left no space for aid, it also left the children of alcoholics untarnished: filial germ plasm would not be damaged by parents' vices. Although alcoholism could be the product of defective stock, in the case where it was not, and the stock was healthy, the germ plasm, and therefore the potential, of future generations remained both intact and guarded against toxic environmental influence. Where Saleeby, in deterministic fashion, saw doom, Pearson saw a new start. Leftist hard-hereditarian eugenicists made similar arguments. As Julian Huxley noted in 1949, Mendelism "makes it clear that even after long-continued bad conditions, an enormous reserve of good genetic potentiality can still be ready to blossom into actuality as soon as improved conditions provide an opportunity" (1949[8]). The argument can be originally found in Alfred Russel Wallace (1892) when he claimed that it was a "relief" that Weismann was right since this implied that "all this evil and degradation [of our present social arrangement] will leave no permanent effects whenever a more rational and more elevating system of social organization is brought about". As we shall see soon, this *topos* had a major influence in Soviet debates on the politics of heredity.[9]

Ernest William MacBride: the inhumane Lamarckian

The ambiguities and slippery terrain of the alcohol debate, especially Saleeby's nurturist position, dissolve in the unequivocally racist positions endorsed by the embryologist Ernest William MacBride.

Professor of zoology first at McGill University in Canada and then at Imperial College London, MacBride (1866–1940) was the most famous neo-Lamarckian biologist in the British Eugenics Society. He strongly opposed the materialism he found in Darwinian views; Lamarckism seemed to him the basis of a less mechanistic view of evolution and eugenics (Bowler, 1983, 1984). He was also a scientific supporter of Paul Kammerer in spite of the radically opposed political implications they drew from their work. In the last years of his life, MacBride expressed Nazi sympathies.

It is therefore easy to be cynical about Arthur Koestler's description of MacBride as the "Irishman with a heart of gold" (Bowler, 1984). MacBride in fact had little sympathy for lower social classes or his fellow Irishmen. As an Ulster Protestant and a political conservative, he was a staunch exponent of scientific racism, always ready to denigrate non-Nordic populations and oppose integration between races. He was a key figure in the renaissance around 1920 of "racial hibernophobia," which reframed the Irish question around "racial opposition... between Irish Mediterraneans and Nordic Britons." According to this racist argument, Ireland presented a "more acute and dangerous" case than other countries given that here Nordic and Mediterranean "confronted each other directly without, as elsewhere, being separated by Alpines" (Douglas, 2002).

Nor did MacBride's heart of gold shine when he advocated extreme eugenic measures, including compulsory sterilization "for the economic sin of producing more children" than one was able to afford (MacBride, 1930). He even celebrated, in extremely Malthusian terms, the fact that the only prosperous district in India had been hit by an inundation that "had drowned 700,000 people in twenty minutes" (1926).

Far from an obstacle to MacBride's racist rhetoric, Lamarckism was one of its pillars. He relied on Lamarckism to justify racism and classism. Races "acquired their characters as a reaction to their different environments" (1924: 241), MacBride claimed, and environmental history deposited in them different biological memories. These memories were "the acquired characters, or better, the acquired reactions of the organism, slowly gained in its struggle with the environment". Modern racial typologies resulted from the accumulation of these different memories. The Nordic race, for instance, learnt its "indomitable courage" in the struggle against the "bleak climate of their old home" (1924: 243). Conversely, the Mediterranean race, formed in a less invigorating climate, was "characterized by a mercurial temperament, prone to quarrel and quick to take revenge.... When they drift into the town they tend to form the 'submerged tenth' i.e. the inhabitants of the slum." The Negro, of course, was "a thoroughly tropical animal" (1924: 242, 244–245).

MacBride even claimed that since racial distinctions were acquired in a Lamarckian rather than a Mendelian fashion, they become harder than hard heredity. He argued that whereas Mendelian mutations under certain circumstances could revert to a normal type, this could not happen with acquired characteristics:

> Racial distinctions...are the most deepseated differences which divide mankind; racial characters are the epitome of a long evolutional history extending over tens of thousands of years – they are the embodiment of a whole hierarchy of memories, disposition, and traditions. (1927)

At least when it came to the past, soft heredity was more stable than hard heredity. When it came to the present, though, heredity was soft and unstable, open to environmental influences – but, again, only pathogenic ones. Here the rhetoric of racial poison furnished supplementary arguments for MacBride's racist and classist discourse. "In addition to racial differences" MacBride wrote, "we have to consider the effects of germ-weakening" which "seems to arise most readily in large towns under conditions which favor overcrowding and unhealthy conditions during conception" (1924: 247). Not only was the past germ plasm of inferior races porous to the enervating conditions of their original environment – as in the case of the Irish – but it was still worsened in modern times by the pathogenic influences of degraded environments where these populations tended to live. MacBride cited the "the slums of Liverpool and Glasgow," filled "with a stunted population of so-called Celts from Wales and Ireland, really belonging to the Mediterranean race" that "breed like rabbits" (1929, quoted in Douglas, 2002).

The double standard by which MacBride considered negative traits always acquirable and positive ones beyond transmission was the clearest mark of his right-wing Lamarckism. If Lamarckism can be opened to a dialectics of degeneration and regeneration, regeneration never emerged in MacBride. The relationship between past and present was completely asymmetric in favor of the former:

> The tacit assumption of philanthropists all through the eighteenth and nineteenth centuries was that the differences between men were due to circumstances and could be abolished by education. Whilst in the last resort racial differences are due to circumstances operating through thousands of years, yet the idea that education and environment *acting through one or two generations can cancel the work of thousands of generations is singularly futile.* The inferior races can be trained in civilized habit and kept in them so long as the superior race is in control, but left to themselves they revert to the stage in development appropriate to their inborn psychic equipment. (1924: 245–246; my italics)

As this passage demonstrates, MacBride's Lamarckism came to resemble hard-hereditarianism. As Peter Bowler writes:

> By insisting that the effect could only work over many generations he had converted Lamarckism into a hereditarian philosophy of nature, at least for all practical purposes. New habits might direct the long-range course of evolution, but one could not expect them to have any immediate effect in altering the inherited character of the organism. (1984: 249)

The tendency of Lamarckism to creep into something like the fatalism of hard heredity was not unique to MacBride, but it was particularly acute in his case. His comfort with hard heredity – though we must stress that he was not a hard-hereditarian, but merely acted like one by asserting the power of an inherited and essentially unalterable biological memory – was certainly singular for a Lamarckian.

In the last years of his life, MacBride's positions became ever more extreme. MacBride trained in Germany in his youth and, according to Pauline Mazumdar, maintained intellectual links with German biological and political thought, which may explain why he found Nazism appealing (Mazumdar, 1992; Bowler, 1984). He resigned from the Eugenics Society in 1931, apparently over an editorial row, (Hodson, 1988) but his extremist politics may have been the underlying cause (Bowler, 1984).

In 1930 he wrote in *Nature* of forced sterilization and the following year, again in *Nature*, he railed against "humanitarian sentiment acting in ignorance of the laws of biology" as a "most dangerous thing" producing "devastating results" (MacBride, 1931). In 1936, yet again in *Nature*, he asked rhetorically why we should "preserve as progenitors of the next generation people who – morally, mentally, and physically – are just as the deformed chamois and deer" (MacBride, 1936). The article was so scandalous – especially considering the political situation in Germany – that biochemist and historian Joseph Needham sent a letter to the editor protesting doctrines "so dangerous to humanity, receiving the imprimatur of what is perhaps the most famous scientific weekly in the world." Was MacBride serious, Needham asked, in suggesting "punishment for the two million unemployed?" (1936) MacBride also wrote letters to the *Times* supporting the brutality of Franco's troops (1937a, b).

Any suggestion that Lamarckism distinguished itself from hard heredity by means of politics, that the former was necessarily an entity of the left, must ultimately founder against the shores not only of Saleeby's sins-of-the-father illiberalism, but also of MacBride's unrepentant racism and classism.

Left Lamarckians: productive eugenics, environmental vitalism, and organic progress

That Lamarckism maintains a left-wing aura in spite of association with the likes of MacBride owes much to the Austrian-Jewish biologist Paul Kammerer (1880–1926), an indispensable figure of left-Lamarckian eugenics and of our quadrant. His scientific story, disgrace, and suicide permanently scarred the image of Lamarckism in the West, even among sympathizers. Only Lysenkoism brought greater devastation to the perception of Lamarckism outside the Soviet Union. For some, such as Koestler and the Soviet makers of the 1928 film *Salamandra*, Kammerer was a persecuted hero. For others, he was unscrupulous and manipulative, a womanizer, and an unreliable scientist who disregarded, consciously or unconsciously, the results of his experiments in favor of his ideological presuppositions. Alma Mahler, his assistant-turned-lover, reputedly said, "Somewhat less accurate records with positive results would have pleased him more" (see comments in Gliboff, 2006; Weissmann, 2010).

Kammerer's life epitomizes many of the ideological battles of his time: Mendelism versus Lamarckism, socialism versus conservatism, the struggle against anti-Semitism – conflicts that, in the end, became unbearable for this creative and unconventional scientist.

After quickly summarizing Kammerer's scientific contribution and cultural significance in his own time – the critical literature (Bowler, 1983; Gliboff, 2005, 2006), including a recent scientific rehabilitation via epigenetics, (Vargas, 2009; Pennisi, 2009; see a response in Wagner, 2009; Weissman, 2010; Gliboff, 2011) is vast and doesn't need to be retold here (see also Koestler, 1971 and Logan, 2013 for a more sympathetic treatment) – I want to focus instead on what historian Mark B. Adams has called the "last chapter" of biologists' writings: the eugenic speculations that typically come after they have made their scientific arguments. Kammerer's soft-hereditarian eugenics show one of the lost tracks in the relationship between biology and society. His "productive eugenics" – which he also called "environmental vitalism" and "organic progress" – are a peculiar and mostly forgotten politics of heredity.

Kammerer believed in the productive power of the environment. However, his theory differed from many post-1945 leftist appropriations of environmentalism because he saw in environmental change not an egalitarian measure to be taken starting from scratch for each generation but something that could be passed on and "enter into the life-sap of generations" (quoted in Koestler, 1971: 28) – an *acquired* characteristic. The term "productive eugenics" (1924; see Logan, 2013: chapter 5) has to be understood as more than a criticism of the mainstream selectionist eugenics that could only change the relative distribution of good and bad genes without creating new traits. Rather, Kammerer's eugenics was productive because it could transform political action into an "organic technology," (*organische und soziale Technik*, 1920) which could leave a positive legacy for future generations. This progress became therefore an "organic progress," which touched and modified the individual body, not just the *Kultur* or civilization:

> All progressive measures, at home and in school, private and public welfare endeavors, education, administration and government, are endowed with a new and more far-reaching importance when dealing with the theory of the inheritance of acquired characteristics. Only then all these institutions serve not only the fleeting moment and the individual, but also eternity and the generation. No wonder, therefore, that everybody professing reactionary tendencies, in private and public life, fervently combats the contention that personally acquired characteristics somehow, and at some time, can be transmitted... If acquired characteristics cannot be passed on... then no true organic progress is possible. Man lives and suffers in vain. Whatever he might have acquired in the course of his lifetime dies with him. (1924: 17–18, 30)

This was the politico-philosophical tenor of Kammerer's science and of the debates around left neo-Lamarckism in the 1920s. Politics was everywhere in that battle, fought over the mutated protuberances of terrestrial-turned-aquatic toads and the acquired coloration of skilfully trained salamanders.

Kammerer's sociocultural context

Kammerer was a product of the Institute for Experimental Biology in Vienna. Known as the Vivarium, the Institute was often seen as "a snotty, Jewish institution" (Gliboff, 2005a), a place where liberal and alternative views of biology were promoted. The head of the zoology department and cofounder of the Institute was Hans Przibram, scion of a rich Jewish family and a famous experimentalist who starved to death with his wife in a Nazi concentration camp in 1944 (Deichmann, 1996).

In the first Austrian Republic – the independent Austrian state of the interwar years – scientists such Eugen Steinach, Julius Tandler, and Kammerer himself pioneered a malleable and developmental view of biology (Logan, 2013). These scientists understood heredity as flexible, shaped directly by the environment via the mediating function of the hormones: heredity was "soft wax in our hands," "just like soft clay in the hands of the modeler or hard steel in the hands of the machinist" (Kammerer, 1920, see Gliboff, 2005). These Austrian biologists worked to fuse endocrinology and Lamarckian heredity in order to destabilize essentialist categories and boundaries – race and gender in particular.

Rejuvenation and regeneration would promote biological transformation of humanity. These were not only abstract ideas but also a range of invasive techniques. Steinach's methods for rejuvenation included testicular transplantation, vasoligatation, and vasectomy. Kammerer suggested using testicular implants to influence the sex drive in homosexuals and hermaphrodites and administering mild radiation to women's ovaries to increase their capacity for breastfeeding (1920).

Rather than emphasize degenerative processes and harmful environmental influences on the germ plasm as right-wing Lamarckians did, the Austrian group sought to revive humankind in both a literal and a more general, utopian sense. Against the destructive effects of the *mechanische Technik* that he witnessed with horror during World War I (Gliboff, 2006), Kammerer sought a utopian and productive *Lebenstechnik* "aimed to provoke those qualities that are of a higher development and advantageous to the individuals and their offspring." His goal was not only to eliminate sickness but also to improve humanity such that even a state of health would mean something greater.

Kammerer's experiments

Kammerer enjoyed a reputation as an exceptionally gifted scientist who primarily worked with reptiles and amphibians. Many of his experiments on the inheritance of acquired characteristics were conducted with toads, salamanders, and newts. His best-known experiment, with the midwife toad (*Alytes obstetricans*), precipitated the 1926 scandal that destroyed his career and likely triggered his suicide.

Kammerer wanted to see if he could induce midwife toads to change their reproductive habits. The toads mate on land, but Kammerer tried to force them to reproduce in water. Not many progeny survived, but the few that did displayed altered behavior. The sixth and final generation of the experimental toads took to breeding in water. But changes came even earlier. After the third generation, a novel morphological structure developed: "a rough, blackish nuptial pad on their fingers and forearm" that helped the male seize the female in the slippery aquatic environment (1924: 53). A new adaptive characteristic seemed to have been acquired. Kammerer found other new characteristics arising in the "modified" toads: enlarged body, stronger arm muscles, and convergence of the forelimbs, all seemingly proof of "functional adaptability" (1924).

Kammerer's results were challenged from many theoretical points of view. The first critical point had earlier been raised by Weismann – atavism. Since toads are originally a water-breeding animal, the appearance of the nuptial pad could simply have been a reversion to the animal's original state, not the inheritance of a new acquired characteristic. The objection was as much scientific as political: Kammerer had shown a retrogression, not a progression toward a new adaptive trait. Other critics argued that Kammerer actually had performed a "Darwinian experiment" by exerting selective pressures on the few surviving eggs (see a reiteration of the critique in Gould, 1980: 81).

But the real trouble for Kammerer was not conceptual disagreement. In 1926, G. K. Noble, a zoologist from the American Museum of Natural History, went to Vienna, where he studied Kammerer's only remaining specimen of the modified midwife toad (original experiments were conducted two decades before). What Noble found was shocking: the nuptial pad was a forgery, the product of an India ink injection. A few weeks later, Noble exposed Kammerer in *Nature*. "The only one of Kammerer's modified specimen of *Alytes* now in existence lacks all traces of nuptial pads," Noble wrote. "Whether or not the specimen ever possessed them is a matter for conjecture" (1926).

The effects were devastating. Many of Kammerer's enemies, including major figures in genetics such as Lenz in Germany and Bateson in England, were very pleased. Some of his sympathizers, such as Jennings, Przibram, and MacBride, were either put off or tried to defend Kammerer's cause. Kammerer himself, who was not present during Noble's visit, did not dispute the doctoring of the specimen but claimed he was not involved (for an alternative explanation, see Gliboff, 2005). A "great friend of Soviet Russia" (quoted in Babkov, 2013: 526), amid the scandal, Kammerer turned down a recently offered appointment in Soviet Union. A few weeks after the publication of the *Nature* article, he shot himself in a remote area of northern Austria.

However, the midwife toad experiment was not Kammerer's only significant one. From salamanders he was able to obtain an acquired change in reproductive habits and coloration. And he actually considered his work with *Ciona intestinalis*, a tunicate or sea squirt, his "deciding experiment" (1924) on the inheritance of acquired characteristics. He amputated the sea squirt's siphons and found "new siphon tubes became longer upon regeneration than the original tubes and that the complete elongation was inherited by the next generation." It was not, Kammerer claimed, the regeneration that was inherited but rather "a locally increased intensity of growth." With his typical flourish, Kammerer concluded:

> The long-siphoned sea-squirts with regenerated germ-plasm give birth to a progeny also long-siphoned. I hope that in this way I succeeded

> in cutting the Gordian knot, not like Alexander the Great with a sword, but with a dissecting knife. (1924: 126)

In his work on *Ciona*, Kammerer tried to reconcile Mendelism and Lamarckism (on Kammerer as an *Alt-Darwinist*, the old-Darwinian school before the rise of Weismann's polarizing views, see Gliboff, 2011). He claimed that only new characteristics, which possess "a great radiating power," could pass from the soma to the germ plasm and thus could become heritable (1924: 105). With respect to previously held traits, instead the standard Mendelian rule applied.

His experiments with the blind cave-dwelling salamander *Proteus* were also significant, considering their policy implications and the media coverage they received. Kammerer claimed to have redeveloped the sight of this *blind* animal through alternating exposure to red light and daylight. "The atrophied visual organ of the blind and bleached newt *Proteus*" was developed "into a seeing eye" (1924: 39), literally resuscitated from its degeneration. The experiment raised plenty of objections, to which Kammerer replied, again in his characteristic style, "The blindness of the newt is nothing compared to the blindness of those 'who will not see'" (1924: 177).

The political message of all these experiments was clear, and, although long-since rejected by most scientists (see Bateson, 1913; Cock and Forsdyke, 2008), well-received by newspapers. The *New York Times* celebrated Kammerer during his lecturing trip to New York as "Darwin's successor" (New York Times, 1923). Regeneration was possible, human-directed evolution was no longer unthinkable, species could be altered, old defects could be fixed, and improvements could be inherited so that progress would not disappear with each new generation. Eugenics need not limit itself to selecting the fit; a "race of supermen" (*Daily Express*, 1923) could be actively bred. Kammerer was happy to contribute to the sensationalism, assuring the *Times* that "the next generation of Americans will be born without any desire for liquor if the prohibition law is continued and strictly enforced" (New York Times, 1923; see Weissman, 2010; Gliboff, 2006, 2011; Logan, 2013).

Kammerer the eugenicist

Like Saleeby and the Soviet eugenicists, whom we will explore next, Kammerer wanted to build a new kind of eugenics, one that differed from the European and North American mainstream. In order to do so, he first had to show that mainline eugenics – with its aggressive rhetoric, fantasies of degeneration, delusion of "superior races," and rejection of "all measures for human welfare, such as medicine and hygiene" as "obstacle[s] to progress" (1924: 261–262) – was an adulteration of Darwinism. For Kammerer, "genuine Darwinism," much like socialism, was a doctrine not of degeneration but of "upward development":

> A reaction among Darwin's successors has striven arbitrarily to separate the last part of this simple and evident doctrine and to give it out as the whole of Darwinism. If the struggle for existence is praised as a progressive principle, the weeding-out by selection as a productive instrument, then Darwinism becomes anti-Darwinism and the theory of evolution a doctrine of retrogression. (1924)

Kammerer believed than an emphasis on the inheritance of acquired traits would challenge this "caricature of Darwinism." Here he sailed the same sea as other Lamarckians who wanted to put the environment first, but, unlike Saleeby or MacBride, he used the primacy of the environment to destabilize racialist assumptions. Race was not a matter of eternal entities or "irreconcilable differences." Environment and development meant that race was a contingent fact – it had changed in the past and could change again in the near future. "Newborn negro children have skins that are no darker than the adult Italians' or Greeks', but become dark in a few weeks," he wrote. "Egyptian babies require three years to develop the typical skin coloring (1924: 266).

There were historical and developmental dynamics in race formation, and, with an eye toward the future rather than the past, this meant that modern environments may efface "the distinguishing marks of race." Common environments could make different races homogenous (268, 279). Mirroring Franz Boas's 1910 study showing the remarkable plasticity of immigrants' skulls since their arrival in the United States, Kammerer claimed:

> Not only are racial marks that show on the surface of the body (such as the color of the skin, eyes, and hair) capable of adaptation, and of being transmitted under different climatic conditions, but even the skeleton may be influenced by changes in the environment. (1924: 269)

Thus Kammerer reversed the meaning right-wing Lamarckians ascribed to race, constructed by environmental influences of past generations. For right-wing Lamarckians, the weight of past experience, which in a sense became if not eternal as right Mendelians would like, at least very difficult to neutralize, hampered any further change in present races.

But for Kammerer, past changes were evidence that further changes could occur in the future. The American type, bred in the melting pot, exemplified such a "new creation."

Because human nature and human heredity could be made and remade, the eugenicist could aim for "the perfection of mankind" (1924: 283). Cooperation, not competition, would join the inheritance of acquired characteristics, and these two powers would advance the organic regeneration of humankind. In visionary, utopian style, Kammerer wrote:

> We are going to avail ourselves of a power (the inheritance of acquired characteristics) for the purpose of regeneration and the acquisition of some other power (the urge for mutual aid lying dormant) that, in the course of generations, they may bring about regenerative results.... There is an inheritable regeneration: we may, we can, we must apply it! (1924: 283–284)

This confidence in powers of regeneration was the distinguishing mark of left Lamarckism. Whereas right-wing – mainstream – Mendelians insisted on the permanency of pure racial lines, and right-wing Lamarckians adopted an "asymmetry" (Logan, 2013: 93) between negative and positive influences, Kammerer's flexible heredity relied upon "the enormous regenerative power of the environment." For Kammerer, the environment was fully loaded with progressive egalitarian values: "Whatever resulted from harmful environment, and was passed on, must be possible of elimination by changing the environment for the better" (1924: 297). Heredity was freed from any limiting force. Biology meant a continuous openness to future opportunities:

> Man has still to acquire sovereignty over the living matter of his own self... Man must acquire the faculty to mold his pliable body and brains according to his constructive urges. (Kammerer, 1924: 284)

In this organic view, which today we would call biosocial, education was the key to achieving "hereditary regeneration." Again education was not a cultural phenomenon but a vital technology, a sort of organ. It amounted, Kammerer wrote, "to one of the most striking organo-technical means to achieve valuable regenerative characteristics, among them the instinct for mutual help and the instinctive yearning for peace" (1924: 285).

It is important to understand Kammerer's theory not as a departure from eugenics but as an alternative form of it. Kammerer was as much a eugenicist as Davenport and Pearson. He was not opposed to the negative policies of mainstream eugenics and their inhumane language. "Surely the elimination of the harmful is an aim worthy of the labor of the very best," he wrote. Humanity could obtain genius through liberation from "the ballast of inferior elements" (1924: 344). But we could do more than merely toy with existing traits, as mainstream eugenicists sought to do:

> We are in the happy position to reach for a much farther goal. Moreover, many a bad trait may be eliminated, not only by prophylactic measures and by suppression, but also by changing it for the better. It would really be a pity if it were otherwise, because ever so many good tendencies always dwell right close to the bad ones. Where there is a superabundance of light, the shadows are the darkest. (1924: 298)

Left Mendelians

The short-lived but hugely interesting life of Soviet eugenics is perhaps the best example of another lost track in political biology: left Mendelism, which occupies the bottom-left box of our quadrant.

By left Mendelism I do not mean the views of leftist scientists who were also Mendelians. This attitude is common in the history of genetics, where not only liberal views but also overtly leftist and even Marxist political convictions have easily cohabited with high-quality work in genetics. Left Mendelism refers to a direct and unambiguous commitment to a eugenic project with the goal of socialist transformation. This transference from science to politics, as in all positions on the quadrant, is based on a strong view of social engineering and on a utopian and future-oriented goal of controlling human evolution and turning it toward social progress, all elements typical of pre-1945 eugenics. There is no intermediate step from the science to the politics, (a reason for which I don't think Kevles' well-established category of "reform eugenics" (1985) for many of the authors here explored is appropriate to express their political radicalism and faith in the eugenic credo[10]).

This was the principal position of Soviet eugenics until its final destruction in the mid-1930s and of H. J. Muller in particular. Hermann Hermanovich, as he was called in the Soviet Union, where he lived from 1932 to 1936, had a direct influence on the construction of Soviet eugenics. He was also a tragic witness to, and inadvertent accomplice in, its ruin (burned by this event, Muller's post-1945 positions, although remaining profoundly eugenic, became increasingly anti-communist and disconnected from the leftist platform he tried to build during his Soviet years, as we shall see).

Both Soviet eugenics and Muller's pre-1945 position are enormously important in revealing the conceptual dissociation between leftism and Lamarckism (the "Lamarxism" hypothesis) and therefore also, symmetrically, between hard-hereditarian eugenics and conservative politics. Left Mendelians broke both of these stereotypes. They argued that the germ plasm, rather than the environment, was "the most important thing that we possess" (quoted in Babkov, 2013: 608), and, at the same time, a sign of egalitarian and revolutionary values. Left Mendelians converged with left Lamarckians in claiming to use evolution for a socialist transformation of society, but their hope lied in the generative power of hard heredity to create a new man, and flourish fully when the disruptive impact of an unequal society would be overcome in socialism.

Soviet eugenics

The Soviet Union became a vibrant center of research in genetics during a time when few countries could boast comparable programs, and this research was often geared toward eugenics. As Loren Graham writes:

> Observers of its early history are frequently surprised to learn that Soviet Russia in the 1920s possessed a strong eugenics movement. One might have expected revolutionary Russia, which prided itself on opposition to capitalist culture and aristocratic privilege, to have stood aside from the movement for "race betterment" which swept the world in those years and led to the establishment of eugenics societies in dozens of countries. To arrive at this conclusion, however, is to carry back into the third decade of this century ideas both about eugenics and Soviet views of man which took clear form only in later years. (1977)

We must therefore excavate the truth of Soviet eugenics. Doing so gives us the opportunity to highlight the turbulence that preceded the seemingly natural alignment between scientific concepts and political values.

Soviet *evgenika* was born in research institutions lead by some of the most distinguished geneticists of the time, particularly Nikolai Koltsov (1872–1940) in Moscow and Yuri Filipchenko (1882–1930) in Petrograd (Adams, 1989, 1990). What had been a fragmented and insignificant movement in Imperial Russia was bolstered and rapidly institutionalized by the rise of Bolshevism (Krementsov, 2011). The 1917 Revolution accelerated the formation not only of genetics research but also of a true eugenics movement. Despite the outbreak of civil war, eugenics "entered teaching curricula and found a grassroots following in the new Soviet Russia" (Kremenstov, 2010). Several Soviet commissars supported the movement: "the Commissariat of Internal Affairs (a police organization) formally accepted the charter of the Russian Eugenics Society; and the Russian Eugenics Society received a small state subsidy" (Graham, 1981).

Institutionalized Soviet eugenics began in 1919 at the State Museum of Social Hygiene (Krementsov, 2010). Two years later, the Russian Eugenics Society was founded "to unite people who do scientific research in eugenics and racial hygiene in the Russian Soviet Federative Socialist Republic" (Babkov, 2013). The *Russian Eugenics Journal* and the *Bulletin of the Bureau of Eugenics*, the official paper of the Bureau of Eugenics of the Academy of Sciences, were both launched in 1922 (Graham, 1981). It was a "Bolshevik eugenics by real Bolsheviks" (Adams, 1990: 154).

The ideological affinities between eugenics and Bolshevism only seem alien in the eyes of the present, after the crystallization of political-epistemological values. In fact, eugenics conformed with the Soviet preference for scientific social planning and shaped the imaginary of several Soviet Utopias and science fiction of the 1920s (for instance Nikolsky's *In a Thousand Years*, 1927, see Stites, 1989). As Krementsov writes:

> Eugenic ideas of "bettering humankind" resonated strongly with the Bolsheviks' early visions of the country's (and ultimately the world's) future: it is telling that [Nikolai] Semashko and [Grigorii] Kaminskii, both commissars of public health, supported eugenics. Like eugenicists, the Bolsheviks believed in social progress and in the ability of humans to direct it. This congruence of interests allowed Soviet eugenicists quickly to institutionalize their field in post-revolutionary Russia. Eugenics provided an array of meanings, which helped the two groups to develop a common language and to foster the dialogue. This shared language also allowed eugenicists to translate their own, often quite esoteric, interests into a language understood and appreciated by their patrons. (2010)

Koltsov offered one of the first sketches for a global eugenics program in post-revolutionary Russia. He divided the discipline into three strands: "anthropogenetics," by which he meant a pure science of human heredity (see also Serebrovsky); "anthropotechnique," the applied dimension of anthropogenetics; and "eugenic religion," which conveyed its most ideological aspects (Krementsov, 2010). Other figures such as Aleksandr Serebrovskii, Nikolai Vavilov, Isaak Agol, Solomon Levit, and Mikhail Volotskoi would soon become leading representatives of the movement. The recent English translation of Babkov's anthology of early Soviet work in human genetics is invaluable in accessing these sources.

The first *evgenika* publications included translations of Thomas Hunt Morgan; historical overviews of eugenics focused on the Russian roots of the movement; family trees and pedigrees of distinguished scientific and literary figures such as Maxim Gorky and Leo Tolstoy; articles on the impact of culture on selection; "pathographies" of military leaders, inventors, and artists; and even a hereditary analysis of the Decembrist uprising of 1825.

The scientific quality of this material is varied, but it is all politically interesting. Soviet eugenics departed significantly from many of the stereotypes that were typical of Western right-wing eugenics and racial hygiene. The emphasis in *evgenika* is mostly on positive measures: from "enhancing the birthrate of the intellectually strongest groups" to arguments for the "removal of all legal, economic, and ideological barriers that impede the movement of people from various classes of society to the intelligentsia" (quoted in Babkov, 2013). The literature is future-oriented, highlighting the human capacity to take control of evolution and favoring the emergence of an active hereditary type, or, in what Koltsov's coinage, *homo sapiens explorans*. Hard heredity here does not mean enslavement to a genetic past but the capacity to shape the genetic future.

Significantly, when analyzing controversial issues such as inborn criminality or Jewish heredity, the goal of Soviet genetics was to dispel stereotypes by, for instance, showing the lack of any significant behavioral difference between Jews and non-Jews and pointing, when necessary, to the appalling social conditions of Jews. Hierarchical discourses of racial and social superiority and inferiority were never on the Soviet eugenicist agenda (Adams, Allen and Weiss, 2005). Of course, it is fair to ask whether this reflected mere accommodation to the new ideological setting (Adams, 1990) or whether these were genuine attempts to create a socialist route to eugenics. Perhaps both accommodation and sincere ideology were at work.

Not that every work of Soviet eugenics avoided controversy. Some ideas would soon become the target of political attacks. For instance, compulsory sterilization laws on the lines of American legislation were occasionally proposed. Even worse was a passage by Koltsov in the *Russian Journal of Eugenics*. He noted that, like American slave-owners, "Russian (landed) proprietors... who had power over the marriage of their serfs and slaves" could have practiced genetic experiments "if the laws of Mendel had been discovered a century earlier." After Hitler came to power, these statements were used to claim that genetics was a "maid-servant for the department of Goebbels" (quoted in Babkov, 2013).

It is after 1925 that, Soviet eugenics began to suffer overt ideological attacks (Graham, 1977, 1981; Adams, 1990a; Krementsov, 2010). In public debates and official publications, opponents proclaimed that eugenics was a bourgeois science, that the emphasis on heredity meant a betrayal of Marxism and a denial of the importance of social relationships, that geneticists were trying to smuggle fascist and capitalist stereotypes into the USSR under the cover of science. Geneticists were decried as Mensheviks and idealists who wanted to promote a theory removed from practice. In the minds of these eugenics foes, Lamarckism offered a better alternative.

Dramatic change came in particular with the "Great Break" of 1929, marked by an anti-technocratic spirit and frontal attack on non-political experts (Adams, 1990a and b; Krementsov, 1997, 2010). "The 'Great Break' spelled an end to the role played by professionals as government advisers and experts in all areas of the country's life," Krementszov writes. "That role was now reserved for party bureaucrats and ideologues" (2010).

The situation for *evgenika* became untenable. The Russian Eugenics Society was closed and its publications suspended in 1930. The 1931 *Great Soviet Encyclopedia* (quoted in Babkov, 2013: 519–524) defined eugenics as a reactionary doctrine, a smokescreen for class interest, and a stalking horse for fascism. According to the *Encyclopedia,* social conditions were the most important – if not the only – factor in human development, and it was "completely wrong even to discuss the issue of biological inheritance regarding human behavior" (Babkov, 2013: 522).

The Nazi seizure of power in 1933, the importance of genetics and biology to Nazi ideology, and the emergence of Lysenkoism after 1934 brought further opprobrium to Soviet eugenics, although genetics somehow survived in disguise (Adams, 1990a). The final showdown occurred in December 1936, at the fourth session of the Lenin All-Union Academy of Agricultural Sciences, where Muller and Lysenko clashed. The cancellation of the international genetics meeting in Moscow in 1937 marked Soviet genetics and eugenics' ultimate defeat.

The ideological issue: Filipchenko's attack on the inheritance of acquired traits

At the ideological level, the most serious problem for Soviet geneticists was challenging the Lamarxist equation, the idea that there was "something inherent in a hypothetical Lamarckian theory of heredity that made it a substantial buttress for egalitarian political values" (Graham, 1981).

The debate over Lamarxism was not only intense but, in line with Soviet politics generally, a matter of life and death. Filipchenko, the most distinguished Russian geneticist, launched a frontal assault on the Lamarxist equation by pointing out the many counterintuitive and disturbing social implications of the doctrine of the inheritance of acquired characteristics. In his 1925 *Inheritance of Acquired Traits* (quoted in Babkov, 2013: 529; see previous excerpts in Graham, and Adams), he described the superficial hopes of soft-hereditarian, neo-Lamarckian or proletarian eugenics "to bring new beneficial hereditary features into proletariat and peasants through purely outside influences." He then made a case that became famous and remained at the center of future debates:

> If inheritance of acquired traits exists then obviously all representatives of this class carry traces of those adverse influences that their fathers and grandfathers and remote ancestors experienced for a long time. Because of this, our long-suffering proletariat and peasants should carry many fewer beneficial hereditary traits and genes for valuable specific features than do other classes that lived in especially good conditions. (Filipchenko in Babkov, 2013)

This was a clever move. Filipchenko was pointing out the inconvenient fact that Lamarckian theory implied a "genetic impairment of those social classes whom left-wing social reformers wanted to help" (Graham, 1981). This is exactly the argument right-wing Lamarckians implicitly or explicitly advanced to support their view of social or racial degeneration: the present germ plasm of certain disadvantaged social or racial groups would reflect the "debilitating effects of having lived for centuries under deprived conditions" (ibid.).

Lamarckists felt "insulted" by Filipchenko's argument (Babkov, 2013: 529). Soft-hereditarian eugenicists and anti-Mendelians attempted to rebut the notion of permanent genetic scars inherited by the underclass. Several Marxist scholars tried to develop sophisticated arguments to counter Filipchenko. Some, such as Volotskoi, thought Filipchenko was biased in considering only the transmission of pathogenic traits while ignoring positive ones (see Graham, 1981). Others thought the argument was a fair one, but socialism would reverse bad environmental effects. The disagreement had a global audience; Muller wrote to Julian

Huxley from Moscow to apprise him of developments in the eugenics community there. Even when Lamarckians conceded that, because of the inheritance of acquired traits, "primitive races existing today have inferior genes," they were quick to add, Muller wrote, that "three generations of socialism will so change the genes as to make all races equal" (quoted in Roll-Hansen, 2005: 214).

However, the controversy remained very much open. When, more than a decade after Filipchenko's original paper, Muller repeated it almost word-for-word during the dramatic 1936 debate at the All-Union Academy, his words were deleted from the conference proceedings and "replaced by three inane and meaningless phrases" (Babkov, 2013: 538). Muller's original argument at the debate regarding "the fascist race and class implications of Lamarckism," (letter to Huxley, quoted above) was the following:

> It is completely natural to come to the conclusion that because the proletarians of all countries, and especially colonial ones, for a long time lived in conditions of malnourishment and disease, and had no opportunity for mental work and in fact were slaves, then they should have become in their hereditary potential a biologically inferior group compared with the privileged classes, as related to both physical and mental features. This is true because according to this theory such phenotypic characters should have been to some degree reflected in the reproductive cells, which develop as part of somatic tissues. (quoted in Babkov, 2013: 538)[11]

But Muller had already lost the political struggle to persuade Stalin to come on his side in the hard versus soft heredity battle, as we will now see.

H. J. Muller (1890–1967)

Hermann Joseph Muller was born in New York City in 1890. A pioneer of classical genetics, he was one of the students and collaborators of T. H. Morgan between 1912 and 1915 in the "Fly room" (Morgan's lab) at Columbia University. With Morgan (and A. H. Sturtevant and C. B. Bridges) he co-authored the seminal *The Mechanism of Mendelian Heredity* where the role of the chromosomes as "bearers of the hereditary material" was formulated (Morgan, Sturtevant, Muller and Bridges, 1915; see Allen, 1978). He was awarded a Nobel Prize in 1946 for his work in radiation genetics (mutagenic effects of x-rays on *Drosophila*). But Muller was not only a great scientist. He was also a staunch eugenicist. As Diane Paul notes, both Muller's first and last papers, respectively written at the ages of nineteen and seventy-six, "developed a eugenic argument." His son's middle name was "Eugene" (Paul, 1987). Eugenics was "the leitmotif of Muller's life," according to biographer Elof Axel Carlson (Carlson, 1981: 393; see also Carlson 2009).

However, Muller's eugenics was out of step with that of his fellow American citizens. Against the reactionary trend of American eugenics, Muller wanted to show how genetics "belonged to the political left-wing" (ibid.). In a famous 1933 article titled "The Dominance of Economics over Eugenics," Muller attacked the "several serious contradictions" that the application of eugenics under capitalism implied: "In order to justify the existence of the gross economic and social inequalities... it has been necessary for the apologists of this system to put forward the naïve doctrine that the economically dominant classes, races and individuals are genetically superior." To which he objected:

> On theoretical grounds, in fact, there is at least as much reason for supposing that the dominant classes represent a selection of socially inferior, as of socially superior genetic material. Thus capitalism leads to a false appraisal of the genetic worth of individuals, and of vast groups, which results in entirely mistaken conceptions of eugenic needs. Our economic system, by exalting the acquisition of private profits, regardless of at what expense to others they were obtained, inculcates predatory rather than constructive ideals. In consequence, the ideal set of characteristics which most present-day eugenists and the population at large would set up as a eugenic goal, is far from the type which would be considered most desirable in a well-ordered society. (Muller, 1933)

Also in light of these political-ideological motivations, Muller, after a Guggenheim Fellowship in Berlin, moved to the Soviet Union in 1932, where, on Vavilov's invitation, he became a leading figure at the Institute of Genetics first in Leningrad and then in Moscow (Carlson, 2009). In a controversial 1934 text, (published in Graham, 1972), Muller even argued for the profound symbiosis between the materialistic view of life contained in genetics and the kind of materialism that undergirded Marxism-Leninism.

Even work done in the Drosophila group, Muller wrote, was the result of a direct influence of Marxist thinking (see also his 1929, The Gene as the Basis of Life).

The most visible document of Muller's attempt to align genetics and communism during his time in the Soviet Union is the letter he sent to Stalin in 1936 as an introduction to *Out of the Night*. The letter (in Glad, 2003; see also Adams, 1990a) was a last desperate attempt to persuade the dictator of the natural alliance between eugenics and communism in the hope of launching a vast eugenic program in the Soviet Union as an alternative to the bourgeois-fascist eugenics of the West. With this bold move, he tried to convince Stalin to go with hard-hereditarianism rather than Lysenkoism at a moment when the latter was ascendant.

Addressed to "Dear Comrade Stalin" from a "scientist with confidence in the ultimate Bolshevik triumph throughout all possible spheres of human endeavor," the letter conveyed a profound sense of urgency. After flattering Stalin's "farsighted view and strength in the realistic use of dialectic thought," Muller begged the dictator to consider "a matter of vital importance arising out of my own science – biology, and, in particular, genetics." This matter

> is none less than that of the conscious control of human biological evolution – that is, the control by man of the hereditary material lying at the basis of life in man himself. This is a development which bourgeois society has been quite unable to look squarely in the face. Its evasions and perversions of this matter are to be seen in the futile mouthings about "Eugenics" current in bourgeois "democracies," and the vicious doctrine of "Race Purity" employed by the Nazis as a weapon in class war. These spurious proposals are offered as a substitute for socialism, i.e., as a decoy to mislead and divide workers as well as petit bourgeois. (Muller, in Glad, 2003)

Rather than emphasize, as in the conservative or racist version of eugenics, the idea of genetic limits on human nature and behavior, this leftist version was replete with notions of continuous progress, perfectibility, and "limitless potentialities." Biology, Muller continued,

> has found no evidence in support of the ancient naïve belief that the physical frame of man, or his congenital mental and temperamental equipment and capacities, have reached any final stage, any divinely ordained suitability. They have not yet come near to "perfection," whatever that may be supposed to mean, or to any physical limit of possibility.
>
> Human nature is not immutable, or incapable of improvement, in a genetic [sense] any more than in a social sense. It is no idle fantasy

> that, by a combination of the favorable education and social material advantages which socialism can provide, on the one hand, with the scientific application of genetic knowledge, unhampered by bourgeois social and ideological fetters, on the other hand, it will be possible within only a few generations to bestow the gift even of so-called "genius" upon practically every individual in the population – in fact, to raise all the masses to the level at which now stand our most gifted individuals, those who are helping most to blaze new trails to life. And even this need only to be the beginning. (ibid.)

Through the power of the genes, Muller's view of the left aimed to obtain what it had usually thought achievable only through nurture. To the pro-environment bias of the left, Muller paid only lip service.

> The usual environmental influences that affect the body or mind of the individual, such as education, better nourishment etc., although they are extremely important in their effects on the individual himself, do not result in improvements or in any definite kinds of changes, of the genes within and so the generations following such "treatment" start in with the same capacities as their forefathers. (ibid.)

Since acquired characteristics were not heritable, nurture could do little to achieve the radical social transformation Muller envisioned.

Even considering possible accidents in the eugenics programs – and that "the children stand, on the average, only half-way" in their progress – Muller wrote that in only twenty years, "there should already be very noteworthy results accruing to the benefit of the nation." And if at that time "capitalism still exists beyond our borders," these genetically engineered socialist populations would confer a decisive advantage in the battle for global socialism.

Muller's underlying narrative was that genetics represented the final step in the communist revolution. Whereas Marx had shown how to take control of social evolution by understanding and using the laws of economics, Muller would show how to take control of biological evolution by understanding the laws of genetics. In both cases, the proletariat would benefit. The time had come to take "conscious social control" of the "grand march of biological evolution, which, through a thousand million years, carried life from microbe to man."

Stalin was not persuaded. In fact, he was greatly displeased. The political situation became impossible for Muller, and he was forced out of the country. On Vavilov's advice, he volunteered for the International

Brigades in Spain to show the Soviet regime his loyalty (in order to protect fellow geneticists in the country), and never returned to the USSR. He was considered a communist, and a traitor, in the US (where during his time in Texas had launched, though just for one issue, a socialist newspaper called *The Spark*) and a geneticist-fascist, and a traitor, by Lysenkoists in USSR.

With Muller's defeat, the possibility of an institutionalized leftist, hard-hereditarian eugenics in Europe disappeared. A few such isolated intellectuals in the West, including Haldane, lingered in small-time organizations such as the British Communist Party, but the prospect of policy impact was minimal. Indeed, they had no voice in government.

Muller, as we shall see in Chapter 6, remained a eugenicist all his life, but the political overtone of his agenda changed significantly after 1945. His devastating experience in Soviet Union, the rise of Lysenko as the Communist party line in biology, and the destruction of Soviet genetics produced in Muller a "180° turnaround" (Carlson, 2011). Muller at the peak of the Cold War, and with the authority of his Nobel Prize, started to embrace a very vocal anti-communist rhetoric, as in the "Spitzer affair", an Oregon professor of biology with Lysenkoist sympathies (see de Jong Lambert, 2012). Huxley also maintained a eugenic agenda under the guise of post-war terminology such as population control and transhumanism, as we shall see, but this was far from the kind of socialist transformation that Soviet eugenics had in mind. The use of genetics to make a socialist revolution, the association between hard-hereditarian eugenics and the left (and, conversely, soft heredity and right-wing arguments, from racism to biological inferiority of the proletarians) went missing from the ideological quadrant. Environment, though now without the productive qualities Kammerer yearned for, became the almost exclusive province of the left.

Conclusions: the political quadrant and its implication for the social construction of knowledge

The early twentieth-century history of the politics of heredity demonstrates clearly that biological doctrines do not entail political values. Diane Paul, Peter Bowler, and Loren Graham have all demonstrated as much, and the quadrant I have traced reinforces the point. Although, as Graham writes, there is a shallow sense in which Lamarckism may lean toward social reform and Mendelism toward racialist conservatism, the bottom line for all these historians is that scientific views are malleable in support of one or another political position. As Graham writes:

> Which way the theories would work in a given historical situation depends on the values of the political and scientific authorities who would employ the theories and the associated technology.
>
> In my opinion the present state of our knowledge of human heredity does not allow an abstract answer – that is, an answer apart from reference to available technologies and existing sets of social forces – to the question of whether available theories of the nature of man have in themselves positive or negative value content. (1977)

Arguments about the values inherent in theories of heredity or human nature are necessarily specious. What one in fact sees are values that crystallized only retrospectively. In a sort of selective Darwinian fashion, "the links that we perceive through hindsight are created by the success or failure of each theory within the available social environments" (Bowler, 1984).

But even if there is no preordained link "between a social position and any particular scientific theory," (Bowler, 1984) even if scientific theories derive their sociopolitical value "much more from their relationships to ... external considerations than from anything inherent in the science," (Graham, 1981) scientific vocabularies still frame political positions. On this point, I diverge from and partly revise Graham's, Bowler's, and Paul's contributions.

The Soviet Union saw genuine conflict between left Lamarckism and left Mendelism. Saleeby's views on alcohol (taken as a representative of nurturism though on the non-Mendelian side) were genuinely different from Pearson's (taken as a representative of naturism). In both cases, alternative scientific doctrines had substantial impact on the construction of values. Surely Mendelism and Lamarckism can lean in all political directions, but a left Mendelian has a different worldview and order of priorities than a left Lamarckian. The same is true of right Lamarckians and right Mendelians, for which degeneration meant two different things, pathogenic effects of environment for the first, outbreeding by worse stocks by the latter.

Had Muller or Kammerer been asked by Stalin or another Soviet leader to design and lead a eugenics program, they probably would have promoted different measures – sperm banks in Muller's case, organic technologies of rejuvenation in Kammerer's. Both were leftists, materialists, evolutionary-oriented, and excited by eugenics. But their worldviews were different and the vocabulary of science likely shaped and consolidated this plurality.

The most significant objection to my idea of a differential coloring of politics by science comes from Bowler's analysis of MacBride (1984). For

Bowler, MacBride's case shows not only that that there can be no rigid connection between scientific and social positions – "Many theories can respond adequately to the same social pressure," as he writes (1984: 260) – but more importantly that concerning imperialism, policies toward the unfit and racial hierarchies, a soft-hereditarian as MacBride is indistinguishable from a hard-hereditarian as Pearson (1984: 258). It is important to give a bit of context regarding Bowler's claim. Bowler's criticism is directed against Donald Mackenzie's *Statistics in Britain* (1981), on the strict connection between biometry and eugenics, and against Koestler's naïve assertion of a one-way link between Lamarckism and social reform. I am entirely with Bowler in his criticism of Mackenzie and Koestler, as well as on the lack of historical nuances of Mackenzie's sociology of biometry, and I agree with him on the reactionary potential of Lamarckism, that is far from being a guarantee for social reform (1983, 1984). I disagree, however, with his pessimistic conclusion.

The implication of Bowler's view seems to be that when there are political stakes, ideology trumps science, making scientific views fungible in the service of ideological goals. Science does not bring anything original to the construction of political agendas or alternative philosophies of human nature. There is no specific way to be a Lamarckian racist; one is simply a racist, and Lamarckism or Mendelism becomes a cover for this.

However, looking at MacBride in the context of my political quadrant, I think a fairer conclusion is that he was an anomaly – not because he was a Lamarckian racist and imperialist but because he held those positions in a way that was nearly indistinguishable from those of a racist hard-hereditarian. MacBride's zeal for social selection (even through bombing) and emphasis on the fixity of racial characteristics is not unique among right-wing Lamarckians, but it is rare. Saleeby or Ellwood's passage on the "negro child" or the many examples of medical degenerationism via environmental effects are more representative of a truly soft-hereditarian right-wing (albeit, again, in authors with often a complex sociopolitical profile).

Thus it does not make sense to argue via MacBride that science is irrelevant in the construction of political views, except to provide an ad hoc imprimatur. Bowler concedes that different approaches to eugenics (biometrical, Mendelian or Lamarckian) may at least frame in a different way the social problem confronted, but the effects remain the same. I think that science does more than that to politics. The differences between Muller and Kammerer, between Saleeby and Pearson, show that scientific views shaped politics and influenced policy agendas.

There would be no quadrant of political-epistemological positions if one extrapolated a general case from MacBride. Rather, there would be only a biological left and a biological right in which political ideologies would manipulate fundamentally inert scientific views to their own advantage. But science is not inert in this sort of co-production, as I have tried to show.

One of the key points of my analysis is that contingent historical events, especially in interwar eugenics, produced the specific alignment of science and values we have assumed natural or logical. But if contingent historical events, rather than logical necessity, produced a certain crystallization of values, then things could have been very different, according to the particular scientific theories that were discarded. Had Stalin made Muller, not Lysenko, his official biologist, would we think today of nurture and soft heredity as liberal values? Of Lamarckism (via epigenetics) as a progressive discourse? It is always dangerous to play with counterfactuals, but, had Stalin embraced Muller's *Out of the Night*, and had positive hard-hereditarian eugenics flourished in the Soviet Union, the post-1945 scenario would probably look different. It is likely that the environment would be less equated with egalitarianism, and the Filipchenko-Muller argument about "deprived environment" resulting in "damaged heredity" (Graham, 1981) would strike more chords today, amid the return of epigenetics. In social-science quarters, genetics would not be equated with sociobiological views of a phylogenetic past anchoring human behaviors but would instead be seen as a source of the openness of human nature to engineering. Perhaps, with soft heredity dissociated from the Soviet Union, the rejection of Lamarckian mechanisms in the West would have been less complete. But this is not what happened. We call right Mendelism mainline eugenics because that was the version of eugenics that crystallized. The implications of this crystallization for the post-war eugenics debate were far-reaching.

But now it is time to look at another counter-intuitive implication of hard heredity, this time in debates on the emergence of an autonomous social science. Here the key figure is Alfred Kroeber, who sought to ally anthropology with hard heredity against racist Lamarckism.

Intermezzo I-Kroeber among the left Mendelians[12]

Alfred Kroeber, one of the heroes of American anthropology, was probably the author who best understood the emancipatory possibilities implied by the rise of hard heredity (Kroeber, 1915–1917, 1952; Krönfeldner, 2009; Lock, 2012; Meloni, 2016). It would be wrong to consider him a left Mendelian along with Muller and Filipchenko. However, Kroeber and left Mendelians had a common enemy: those forms of "inheritance by Magic" (Kroeber, 1916b) promoted by Lamarckians, in which both Kroeber and the left Mendelians saw potential for social and racial inequalities. There are, besides, interesting and much neglected biographical overlaps between Alfred Kroeber and a key left Mendelian like Muller. Not only because of their common German ancestry, their lives in New York City (where Kroeber moved a few years after his birth in New Jersey) and the relationship with Columbia University, but because the two, with Kroeber fourteen years senior, were actually first cousins on the side of Kroeber's mother, Johanna Muller Kroeber (Theodora Kroeber, 1970). We can only speculate here if and how this kinship meant also a common intellectual background or similar political experiences. But the disciplinary context and trajectory were certainly very different.

Kroeber was responding to an American anthropology that, before the rise and consolidation of Boasian culturalism, was immersed in a confusion of biology and culture, "race" and "civilization" that could not be untangled without the well-defined concept of heredity supplied by Galton and Weismann, and later Johannsen. According to George Stocking (1968, 2001), in anthropological writings of this period, notions about blood, racial temperament and racial memories, and habits becoming instincts indicated epistemic confusion between the biological and the sociocultural "What was cultural at any point in time could become physical," Stocking writes, and "what was physical might well have been cultural" (Stocking, 2001: 14). Nineteenth-century race was an "accumulation of cultural differences carried somehow in the blood" (2001: 8).

North American and European sociology, heavily influenced by Herbert Spencer, offered similar evolutionary social-cum-biological thinking. For Spencer, Lamarckism played a major theoretical role. It provided a mechanism by which habitual behaviors turn into biological instincts and cultural inheritance is rooted in biological heredity. This vision was so widespread that even sociologists who opposed some of Spencer's political proposals (laissez-faire political economy, for instance) shared his Lamarckian mechanism (Calhoun, 2007; Breslau, 2007).

Kroeber attacked these thinkers who welcomed "alien" explanations of "historical phenomena by organic processes." In doing so, he showed that the rise of hard heredity had crucial implications in distinguishing the social from the life sciences.

The problem with this sort of quasi-Lamarckian social science was, Kroeber claimed, that its "blind and bland shuttling back and forth" between cultural and biological mechanisms ("an equivocal race and an equivocal civilization," Kroeber said, 1952), made disciplinary distinction impossible. If a real autonomy of the sociocultural was to be achieved, the Gordian knot of Lamarckian inheritance had to be cut.

Kroeber saw the folding of the hereditarian substance into the germ plasm as the theoretical condition that would emancipate the social sciences from biological heredity and reserve their role to the study of "purely civilizational and non-organic causes" (1917: 182–183; see Krönfeldner, 2009; Meloni, 2016). Reading his 1917 *The Superorganic* today, one is struck by Kroeber's admiration for Galton – he "has always evoked my complete respect and has been one of the largest intellectual influences on me" Kroeber, 1952: 22) – and, above all, Weismann, in whose writings Kroeber sensed an anticipation of "modern cultural anthropology's argument that man's mind and culture were independent of biology and depended upon tradition and other social processes" (Cravens, 1978: 38).

The Superorganic is actually the culmination of a series of texts in which Kroeber came to terms with the Mendelian revolution as the basis for a new, independent social science. In the "Cause of the Belief in Use Inheritance" (1916a), for instance, Kroeber criticized the "naïve, unscientific, and even primitive method of reasoning by analogy" (1916a: 370) typical of neo-Lamarckian authors. Lamarckism is seen as the infancy of a biological discipline that achieves with Weismann "plain maturity." In this mature phase, "organic phenomena must be interpreted solely by organic processes" (1916a: 369). In "Eighteenth Professions" (1915) and "Inheritance by Magic," (1916b) Kroeber further attacks that "vitiated mixture of history and biology" (Kroeber, 1915: 285) seen in the work of Lamarckian authors by using the Mendelian "utter separateness" between the gamete and the zygote (Kroeber, 1916b: 27) to infer that biology cannot in any way explain the achievements of human society (Lock, 2012). "Belief in acquired heredity is merely a result of the failure to distinguish between social and organic processes, and a remnant of the ruder vision of former times when heir meant both a descendant by reproduction and the inheritor of possessions" (1916b).

Although his view of the superorganic was challenged also in Boasian ranks, Kroeber's legacy was lasting. His strategy, adopted by many other social scientists, was to immunize social facts from biological ones – to "disregard the organic as such and to deal only with the social" (Kroeber, 1952: 34–35). In a very different cultural context, and several years before, Émile Durkheim similarly saw social phenomena as explainable only via social explanations, and in a sense the parallelism between the two are evident (as many have noticed, see Degler, 1991; Lock, 2012).

The "liberation" of the sociocultural from the biological followed different paths in different countries, although many scholars have made more complex and nuanced this sort of progressive fable of a one-way emancipation of sociology from biology (for instance: Gissis, 2002; Krönfeldner, 2009; Renwick, 2012; Meloni, 2016). Kroeber's case may be particularly visible and also idiosyncratic thanks to the strong presence of genetics research in the United States. Unquestionably, though, when a hard-hereditarian view arose in Western countries, the Lamarckian matrix on which much nineteenth-century social science had relied entered a deep crisis. A novel division of labor across disciplines emerged, which each discipline finding its own peculiar way to depart from the biological – psychology, for instance, with behaviorism and psychoanalysis (Cravens, 1978; Richards, 1987; Degler, 1991).

In our age of biosocial investigations and repeated calls to go beyond the nature/nurture dichotomy, it is easy to criticize Kroeber's neat and modernist distinction between the organic and the cultural. However, in his own time, Kroeber felt, and not without reason, that a decoupling of the social from biological heredity was the only way to fight racism and limit the expansion of aggressive eugenics. He believed that decoupling the superorganic from the organic meant people and cultures could change. No one was stuck in their own past, as this past was not fixed. Racial differences were superorganic, as was genius, in the sense of a social product, not an individual mental faculty: "Any population of substantial size contained a range of individuals, among whom one or more was capable of becoming a Mozart, *providing* his social or historical environment was capable of realizing that potentiality," Carl Degler writes, summing up Kroeber's view (Degler, 1991).

Distinguishing two kinds of heredity was unmistakably important in the fight against eugenics as well. Here Thomas Hunt Morgan used a similar strategy. As a sophisticated geneticist Morgan was convinced that pedigrees of criminals, degenerate and cacogenic families, so widespread in American eugenics, confused two categories that, in a move very close to the Kroeberian strategy, had instead to be kept separated. These pedigrees were meaningless to Morgan as they mixed hopelessly "effects transmitted biologically with those transmitted culturally" (Allen, 2001). Morgan wrote in 1925:

> The pedigrees that have been published showing a long history of social misconduct, crime, alcoholism, debauchery, and venereal diseases are open to the same criticism [*i.e.*, conflating biological and social heredity] from a genetic point of view; for it is obvious that these groups of individuals have lived under demoralizing social conditions that might swamp a family of average persons. It is not surprising that, once begun from whatever cause, the effects may be to a large extent *communicated* rather than *inherited*. (quoted in Allen, 2011, 201–202; my italics)

The very possibility of establishing a boundary between cultural communication and biological heredity, environmental and genetic effects, was the great political legacy of progressive hard-hereditarians. Morgan named it "the two-fold method of human inheritance" (in Allen, 2001), which clearly converged with Kroeber's distinction between organic and social evolution.

Kroeber found in Weismann and Galton possibilities that they did not. Removing social phenomena from the nightmare of biological heredity was not on their agendas, and Kroeber criticized both for that reason. They had not, he thought, fully recognized the consequences of their own ideas. Galton's eugenics was an impossible shortcut bridging unbridgeable forms of evolution, another form of confusion that the great boundary-maker of cultural anthropology was ready to dispel.

To Kroeber's credit, as we shall see in Chapter 5, post-war attempts at making race a secular and non-mystical concept, now decoupled entirely from the confusing addition of civilizational aspects, all followed the avenues he (and geneticists like Morgan) opened. Heredity and heritage could now travel on separate tracks.

5
Time for a Repositioning: Political Biology after 1945

The second era of political biology emerged from the ravages of the Second World War. In the period following 1945, one observes *in vivo* the coproduction of new epistemology and sociopolitical values. Coproduction goes both ways: Politics restricted the range of acceptable scientific claims after World War II, and a newly reformulated science of evolution, already running since the late 1930s, reshaped the politics of biology and politics at large.

In the aftermath of World War II and the defeat of the Axis powers, the post-1945 world entered a new global dimension (Hobsbawm, 1995; Calvocoressi, 1991; Reynolds 2001; Barrett and Kurzman, 2004; Iriye, 2014). Two new superpowers, the U.S. and the USSR, soon both with nuclear capacity, replaced old colonial actors with important implications at the ideological level. The legitimation of the right to self-determination (Burke, 2010) forced the displacement of classical racism – the hierarchical view of races and racial development typical of the colonial phase – by more "modern" racial views.

As a consequence of the devastations of World War II and the necessity to establish a novel international order, new economic, financial (such as the IMF and the World Bank, Bretton Woods, 1944), juridical, and political institutions were born. Of particular importance to the development of political biology was the United Nations and its subsidiary, the United Nations Educational, Scientific, and Cultural Organization, UNESCO. As Barrett and Kurzman (2004) write:

> World War II dramatically altered the shape of the world polity. The number of inter-governmental organizations and international nongovernmental organizations doubled or tripled (...). In qualitative terms, the replacement of the League of Nations with the

> United Nations signified the increased importance of the world polity vis-`a-vis national states. Whereas the League had been limited in its membership and its resources, the United Nations aspired to universal membership and built an elaborate bureaucratic machine. The United Nations was far from being a world government, and had few coercive powers of enforcement, but it offered a public forum for the expression of opinions and proposals that individual national states may outlaw. Thus, the growth of the world polity after World War II represented a significant international political opportunity.

These novel institutions embodied a spirit not only of internationalism, but also of universalism (Cassin, 1968) and creation of a universal man (Haraway, 1990). A new global ethos was in the making with the goal of a post-war pacification and a rescue of the lost values of humanity after war's atrocity. Hence in December 1948, the UN adopted the Universal Declaration of Human Rights, recognizing "the inherent dignity and...the equal and inalienable rights of all members of the human family." At least theoretically, there was no place for hierarchical and discriminatory views in this inclusionist and egalitarian ideology of personhood (Barrett and Kurzman, 2004).

The emergence of a new set of global norms (ibid.) necessitated a shift in scientific practice and ideas and a new place for science in human affairs. 1945 was a political and ethical watershed, and it is unthinkable that the same science of human nature that marked the interwar period, with its bold biologism, social engineering, and racist attitudes, could survive intact into the novel scenario. The naïve and bold scientism that dominated the interwar moral economy was now an object of suspicion, at least among more sensitive intellectuals, although popular perceptions remained highly scientistic (especially in the U.S., where technocratic optimism remained widespread, see Taylor 1988). Science – from Nazi experiments to the atomic bomb – could have sinister implications, and all kinds of scientists, physicists or biologists, brought now the weight of a heavier responsibility. As Albert Camus commented two days after Hiroshima, "Any city of average size can be totally razed by a bomb the size of a soccer ball" (quoted in Gordin, 2009). A new perception was of the social meaning of science was taking place.

The immediate post-war years, beginning the second era of political biology, were a time of both consolidation and flux. From the viewpoint of our analysis two factors stand out. Firstly molecular biology, a new highly successful research program sets in motion since the 1930s (Olby, 1974; Kay, 1993) had a key role in closing the dispute between hard and soft hereditarians in favor of the former, ending (at least for a few decades as we shall see) a turbulent dispute – though the firm establishment of hard heredity also benefited from extra-scientific motivations. Secondly, a huge movement of repositioning of biology occurred within the framework of the newly constituted human-right, universalist ethos of post-1945. The democratization of biology, however, was anything but calm, reflecting the absence of a dispositive scientific shift in concepts of race and human nature. For many thinkers, it was Nazi horrors, not new science, that precipitated a change of mind or recasting of now-discredited ideology along more acceptable lines. Some scientists sought to maintain eugenics ideas but to discard the evil associations that had accrued to them. This was no mean feat, given the strong and embarrassing ties between, for instance, American and Nazi biology. Other scholars put forward radical new agendas that exchanged biological notions of variation for cultural ones, asserting that only culture could be a source of human difference. The result was all manner of contradiction, as expressed powerfully in the UNESCO statements on race and the debates surrounding them.

Hard heredity consolidates and becomes molecular

The most important conceptual change in genetics was undoubtedly the completed transition from the Mendelian gene to the molecular one, culminating in the 1953 discovery of the double helical structure of DNA (Watson and Crick, 1953a and b; Olby, 1974; Judson, 1996; Keller, 2000; Griffiths and Stotz, 2013; Rheinberger & Muller Wille, 2015). Molecular biology, a term coined in the late 1930s by mathematician Warren Weaver of the Rockefeller Foundation, was more than a mere research program; it brought with itself a highly reductionist view of biological processes, the emergence of what has been called "the molecular vision of life" centered on DNA and based on "mechanisms of upward causation" as "necessary and sufficient explanations" of how life works (Kay, 1993). However, from the viewpoint of our analysis, none of these profound epistemological changes challenged the process of consolidation and final monopoly of the notion of hard heredity that we saw already prevailing in the second part of our first era of political biology. Quite the opposite. The molecular gene, through the establishment of Francis Crick's central dogma of molecular biology (1958, 1970), put to rest the controversy over hard and soft heredity, as we shall explore more in detail in Chapter 7. For decades, the prevailing trend had favored hard heredity, and DNA sealed now its victory. Crick declared that information flowed only one way, from DNA to RNA, never in reverse. DNA, in other words, is only the source, never the receiver, of biological information (1958). As he restated in 1966 Croonian Lecture on *The genetic code* "the cell can translate in one direction only, from nucleic acid to protein, not from protein to nucleic acid. This hypothesis is known as the Central Dogma". With the Central Dogma in no way the environment can send signals to the genome.

Crick's argument was, in many respects, a molecular duplication of Weismann's barrier (Griesemer, 2002). The anti-Lamarckian dimension of this argument was made explicit, by Crick himself, but also by big names of evolutionary thought such as Ernst Mayr and John Maynard Smith. "The greatest *virtue* of the *central dogma*" Maynard Smith claimed "is that it makes it clear what a Lamarckist must do – he must *disprove* the dogma" (quoted in Gissis and Jablonka, 2011).

More generally in terms of evolutionary theorizations, the rejection of Lamarckian inheritance was one of the distinctive traits of the Modern Synthesis: not only the same soft-hard heredity dichotomy (where soft is the "pre-modern" pole) was a modern synthesis construct (see footnote 1, Chapter 1), but also the synthesis can be described as a process of growing "constriction" (Provine, 2001) or repudiation of the many alternative evolutionary views that used to compete with Neo-Darwinism (Bowler, 2009). Lamarckism and its soft-hereditarian penchant were one of the first to fall in this process of "hardening" of the Synthesis to use Gould's famous term.

However not just the discovery of DNA, but also ideology was a powerful factor of constriction, repudiating alternative evolutionary views that once competed with neo-Darwinism. The Cold-War context in particular was key in further marginalizing soft-heredity.

That soft heredity was propagandized as a state doctrine by Soviet Lysenkoists, especially between 1948 and 1952, was an important factor in consolidating the hegemony of hard heredity in Western science. Historians have noticed that it was common in the 1950s for people working in a slightly unorthodox way on non-Mendelian or cytoplasmic inheritance to be accused of Lysenkoism (see Sapp, 1987, 2003). Tracy Sonneborn was one of such cases. His research on cytoplasmic inheritance in the ciliate protist *Paramecium* was often associated with Lysenko, something especially unhelpful in America in the 1950s. This "vicious rumour" put him on the defensive, pushing him to ally more neatly with nuclear genetics, and drawing clear boundaries between his work and any form of alternative heredity often mistaken for "Lysenkoism" (Sapp 1987). It is also interesting to remark that another unorthodox author, Conrad H. Waddington (one of the fathers of epigenetics, as we shall see in Chapter 7, and proponent of a synthesis of embryology and genetics), had his first Russian translation in the leading Lysenkoist journal, *Agrobiology* (Lambert and Krementsov, 2012) although Waddington publicly attacked Lysenko during the 1950s (see Jones, 1988). Another example (Lambert and Krementsov, 2012) of the stricture in the intellectual climate is the 1961 revision of the entry Lamarckism for the *Encyclopedia Britannica*. Authored by Conway Zirkle, professor of botany at the University of Pennsylvania, who added his name to Morgan's original entry, it now displayed anti-Lysenkoist, Cold-War tones (Zirkle had previously written a book on the destruction of genetics in USSR). Such ideological constrictions limited scientific heresies (Gilbert, 2011; see also Lamb, 2011; Roll-Hansen, 2011), but also impeded the creativity inherent in them. As William deJong-Lambert and Nikolai Krementsov write (2012), it is important to consider

> whether Lysenko's identification with Lamarckism, and the relentless anti-Lamarckian ridicule this provoked, may have impeded the advancement of approaches to the study of biological evolution, which emphasized the role of the environment such as C. H. Waddington's epigenetics.

In sum, for both ideological and scientific reasons, the post-1945 phase of the hard-versus-soft-heredity debate was a period of reduced turbulence courtesy not only of a new molecular language emerging in biology but also because of broader ideological factors. When both these internal and external factors eroded, as we shall see in Chapter 7, the conditions to rethink in a less rigid way the dichotomy of hard and soft heredity were eventually possible.

The democratization of biology: containment of racism and eugenics

Alongside the hardening of hard heredity, the democratization of post-war biology is the second main visible trait of the mutated scenario. As anthropologist Jonathan Marks notices, at the end of the Second World War "the fields of human genetics and physical anthropology were in disrepute and needed to be reinvented". With post-war biology bent toward a liberal-democratic, universalist, and human rights framework (Haraway, 1990), the wild utopian rhetoric of the eugenics heyday had to be tamed. Scientists announced a series of breaks with the past. They adopted a new palimpsest or "script" (Selcer, 2012) with which to negotiate social and political pressures and an elaborate a new "politics of presenting" science to the wider public (Brattain, 2007; Weidman, 2012).

Take race for instance. As Anthony Hazard writes, race had to be reconstituted "because the political exigencies of the immediate post-war period called for anti-racist pronouncements from the political and scientific communities of those nation-states who defeated the purveyors of the Holocaust and whose domestic policies were being broadcast internationally" (2012).

The goal was not only to break with the past, though. Because the United States positioned itself as the defender of free people against communism, it had to reckon anew with their own racism. This latter factor is no less important than the former. Studies like the one by Dudziak (2000) have highlighted the necessity for the U.S. to advance, because of international pressure and the dawn of the Cold War (which started officially in 1947), a new reformist rhetoric "in order to make credible the government's argument about race and democracy" (ibid,: 14, see also Furedi, 1998). To mark a break with the past but also to counter international criticism, new measures had to be taken on inconvenient topics like race. This was the underlying message of Truman's famous "Special Message to the Congress on Civil Rights," delivered in February 1948.

> If we wish to inspire the peoples of the world whose freedom is in jeopardy, if we wish to restore hope to those who have already lost their civil liberties, if we wish to fulfill the promise that is ours, we must correct the remaining imperfections in our practice of democracy. (quoted in Dudziak, 2000)

However, it is important to clarify that the intense boundary-making that brought biological science in line with the exigencies of the post-war world was not purely a result of external pressures. These pressures worked in tandem with a process that had already begun in evolutionary thinking. A series of important transitions, which had commenced in the 1930s but became more visible after Second World War, converged in what is known as the modern synthesis (Huxley, 1942; Provine and Mayr, 1980; Smocovitis, 1992, 1996). In scientific terms the modern synthesis was a process of unification of many strands of biological research, from microevolution to macroevolution, from population genetics to paleontology, a unification under the combination of a mix of genetics and natural selection, Mendelism and Darwinism, now finally reconciled after the rift we have described in Chapter 3 (Fisher, 1930; Dobzhansky, 1937; Mayr 1942; Simpson, 1944; Stebbins, 1950). However, the modern synthesis was also more than a scientific integration of evolutionary thought. It was also a certain ethos perfectly embodied by some of its key figures, as we shall soon see. "A liberal, humanistic, and secular worldview" (Smocovitis, 1996: 99), the modern synthesis was the keystone of the new consensus in biological thought. According to historian of the synthesis Betty Smocovitis:

> As the horrors of the Holocaust became known and nuclear threat loomed with the Cold War, a framework that endorsed the fundamental adaptability of life that offered some progress, a moderate or liberal ideology, and an optimistic and coherent worldview with humans as agents of their own evolution became even more urgent. (Smocovitis, 2009)

The implications for themes like race or eugenics were profound, though not unambiguous. Race did not disappear, but, largely thanks to the architects of the synthesis, it would be reformulated and constrained by well-defined political and moral boundaries. On the one hand, after 1945 overtly racist claims in public evolutionary writings were largely discredited and marginalized within the mainstream liberal consensus (Jackson & Weidman, 2004; Jackson, 2005). Their authors were increasingly defined as "bogus" scientists or "cranks." On the other hand, racial theorizations took "advantage of the logical weaknesses of anti-racism" (Brattain, 2007), which, as we shall explore in detail later in this chapter, continued to rely on race although in principle as a purely technical biological concept with no supposed effect on human behavior. The second era democratized biology; however, as a growing body of scholarship has highlighted (Gannett, 2001, 2007; Reardon, 2005; Abu El-Hay, 2007, Yudell, 2013; Lipphardt, 2014), what occurred was a reconceptualization of race according to a different game of truth. Racism was not eliminated but contained, displaced, and reconceptualized.

The same is true for eugenic attitudes, which did not evaporate overnight after 1945. Muller and Huxley remained consistent and visible eugenicists for their whole lives. As the first director of UNESCO from 1946 to 1948, Huxley called for a renewed eugenic agenda:

> Though it is quite true that any radical eugenic policy will be for many years politically and psychologically impossible, it will be important for UNESCO to see that the eugenic problem is examined with the greatest care, and that the public mind is informed of the issues at stake so that much that now is unthinkable may at least become thinkable.

Huxley continued to write on biological inequality and the dysgenic effects of modern civilization, indicating that "a eugenic value system continued to operate in the social sciences." He, Muller, and others who shared their views "enjoyed immense status and prestige" until the libertarianism and counterculture of the mid-1960s (Weindling, 2012) initiated a radical critique of scientific expertise (though, even at that point, scientists such as Carl Sagan or Linus Pauling continued to propose to the public major radical eugenic measures). But after 1945 Huxley and Muller made major efforts to rebrand their eugenic agendas in new terms. "Mutations load" became one of Muller's preferred phrases; "population control" functioned likewise for Huxley.

But the containment of eugenics wasn't just superficial or cosmetic (Proctor, 1988). Change was meaningful. A new group of authors became public spokesmen for a different style of political biology. Ernst Mayr, Sewall Wright, George Gaylord Simpson, Theodosius Dobzhansky, Leslie Dunn, and Edmund Sinnott, among others, were the changing face of evolutionary thought. None of them was significantly involved with the eugenic past, and some – Dobzhansky, Dunn, and Simpson in particular – were overt critics of past applications of biology to human affairs.

These authors promoted genuinely scientific constructs and spun them such that the new science offered a publicly understandable break from prewar biology. Emphatic claims of theoretical novelty, such as the shift from typological to population thinking, reflected a revision in scientific views and reassured the public that a new truth-game was underway, that the modern synthesis had nothing to do with the despicable eugenic and racialist past.

Compare two seminal works of the modern synthesis, Fisher's 1930 *Genetical Theory of Natural Selection* and Simpson's 1949 *The Meaning of Evolution*. As we have seen, Fisher's thinking was starkly eugenic. The last third of the book is ripe with classical dysgenic concerns such as the fear of differential reproductive rates between the "prosperous" and the "lower" classes. Meanwhile Simpson's post-war discourse no longer allowed room for racial, eugenic, or political ambiguities. Peppered throughout *The Meaning of Evolution* are terse three-word sentences such as "Authoritarianism is wrong" and "Totalitarianism is wrong" (321). "Eugenics," Simpson writes,

> has deservedly been given a bad name by many sober students in recent years because of the prematurity of some eugenical claims and the stupidity of some of its postulates and enthusiasms of what had nearly become a cult. We are also still far too familiar with some of the supposedly eugenical practices of the Nazis and their like. The

> assumption that biological superiority is correlated with color of skin, with religious belief, with social status, or with success in business is imbecile in theory and vicious in practice. (335)

Here Simpson, a major and a very effective writer (Smocovitis, personal communication), reveals the new ethos of post-war biology – its desire to demarcate itself from past bad science – but also its problems. The old science is neatly condemned, yet Simpson also offers three backdoor apologies for eugenics. First, it failed not because eugenics was inherently wrong but because of the "prematurity" of some claims. Second, Nazi policies pretended to be eugenical but were not. And third, "imbecile" ideas were widespread only among "Nazis and their like," a formulation that conveniently downplays the convictions of the overwhelming majority of British and American eugenicists, who not only shared but perpetuated Nazi values. This demarcation of bad Nazi eugenics from good Anglo-American science was a common rhetorical strategy at the time (see Weindling, 2004; see an example in Bajema, 1976).

The fathers of the modern synthesis frequently spoke of the horrors of Nazism as the embodiment of negative eugenics and the source of urgency behind post-eugenic biology (Simpson, 1949; Mayr, 1982; Dunn and Dobzhansky, 1952,). The same was true for the foundational UNESCO statements on race. As Perrin Selcer writes:

> Nearly all historical references to UNESCO's Statements on Race begin, like most contemporary accounts did, with some version of, "In the wake of the horrific mass murder of six million Jews" It would seem hard to exaggerate the transformative power of the Holocaust on the intellectual and political history of race. (2012)

It is true that mainstream thinkers, especially in England, had begun to distance themselves from some of the more extreme misuses of biologism before the Nazi horrors were known. In 1935 Huxley, H. G. Wells, and anthropologist Alfred Court Haddon published *We, Europeans*, the first consistent critique of biological racialism. But this was just one book produced by a divided scientific community (Barkan, 1992: 308). Moreover, it had significant ambiguities (it is enough to look at the Eurocentric title), and Huxley never repudiated eugenics and racism, openly defending British imperialism in 1944 and denigrating the self-governing capacity of Africans whom he dismissed as "rhythm-lovers" (Furedi, 1998). When the UNESCO commission for the race statement was launched, he even suggested Darlington, an overt racist, as an expert for the commission on the race statement, according to Dobzhansky a move to be attributed to Huxley's "rumored senility" (Brattain, 2007: 1400).

As Will Provine has argued in his fine-grained historical analysis of the attitude of geneticists toward race, "Although attacking Nazi race doctrines severely," many of the reform eugenicists of the period from Huxley to Haldane "stopped short of denying that there might be hereditary mental differences between human races or that race mixture held no biological dangers" (1986: 872). According to Provine, "The great majority of geneticists before World War II continued to believe that races differed hereditarily in intelligence, and in particular that African blacks were in a populational sense less intelligent than whites" (ibid.: 873).

In practice, before 1940, critical reflections by biologists on what would become known as the misuses of biological thought were rare. In 1939 C. P. Blacker, secretary of the Eugenics Society, wrote to the demographer Carr-Saunders:

> Whilst understandably disturbed by aspects of German policy...it would be a good thing if the impression were removed that as a Committee we disparage the results of the German policy. For my part I regard these as substantial and indeed remarkable. (quoted in Jones, 1986)

Before 1945, geneticists who felt uncomfortable with eugenics preferred a politics of silence rather than publicly embracing alternative views. Theodosius Dobzhansky, who would eventually emerge as the true embodiment of the new liberal-democratic biology, had already started to publish on populational thinking, one of the key strategies for the post-1945 democratization of biology. However, his *Genetics and the Origin of Species* (1937), though it includes a paragraph on the notion of race from a non-racialist perspective, did not fully engage with human races. "In practice," Gayon writes, "prior to the outbreak of World War Two, Dobzhansky wrote nothing on Man." It was Dunn who, after the beginning of World War II, convinced Dobzhansky that he could take "an explicit antiracist stance" (Gayon, 2003). Dunn later recalled being asked by a radio station to rebut Nazi racial science during the war. He realized that "a lot of these things hadn't been openly discussed, particularly from the standpoint of what the idea of race might be, from the standpoint of the geneticist looking at human races" (quoted in Gayon, 2003).

The ideological transformation of people such as Dunn – from prewar flirtation with eugenics to wartime skeptic to post-war critic – was a complex process, far from univocal (Gormley, 2007, 2009b). However, in spite of the entanglements and ambiguities, it is fair to say that biologists' public statements changed only after the outbreak of World War II, and the radical break came only after the war's end. As Jonathan Marks writes:

> News of the Nazi atrocities repulsed people everywhere. No biologist wanted to give support to Nazi-like race doctrines, including assertions about hereditary mental inequality of races. After the war, only two geneticists, C. D. Darlington in England and R. R. Gates in the United States, made public statements indicating a belief in hereditary mental differences between human races.

Highly visible public statements reflected this newfound awareness. In 1948, the World Medical Association released a post-war version of the Hippocratic Oath, including, *"I will not permit consideration of race,* religion, nationality, party politics, or social standing to intervene between my duty and my patient." Two years later, UNESCO issued its first statement on race, "The Race Question," responding to the Holocaust and the Nuremberg trials (Brattain, 2007; Fullwiley, 2008). As anthropologist and UNESCO rapporteur Ashley Montagu noted, the location of the group's meeting, the former administrative headquarters of Nazi-occupied France, signaled the evils to which the new theory of race was intended to respond:

> Only if our deliberations had taken place at Auschwitz or Dachau could there have been a more fitting environment to impress upon the Committee members the immense significance of their work. (quoted in Hazard, 2012: 38)

Nazi "Barbarous Utopia"

Although a full awareness of the public significance of the Holocaust was a complex and long phenomenon (see Levy and Sznaider, 2004), and the entanglement of Nazi ideology and biological rhetoric was fully documented only since the 1970s, this does not mean that the exceptionality of the Nazi crimes, and their connection with a certain murderous view of science and racism, were not perceived in the immediacy of the end of the war (as Montagu's above quotation illustrates). First reports on human experiments at Auschwitz were known since September 1945, and after 1946 most of the crimes of Nazi doctors at Auschwitz and other camps were of public dominion (Jones, 1988). The same notion of genocide ("the destruction of a nation or an ethnic group") was coined in 1944 by Polish émigré Rafael Lemkin (Cooper, 2008).

Always in 1948, the UN Genocide convention took place, and was ratified three years later in 1951, coordinated by Lemkin himself, who played also a role at the Nuremberg trials.

The notion that the revelations of the crimes of Nazi doctors, the Holocaust, and the public effects of the Nuremberg trials (1945–1946) played a foundational role for the new post-1945 culture has been convincingly claimed for the emergence of bioethics (Annas, 2005, 2010; Annas and Grodin, 1992; Steinfels, 1986; Caplan, 1992) as well as for the new culture of human rights, based on a new universalist and cosmopolitan ethos (Annas and Grodin, 1992; Levy and Sznaider, 2002, 2004; Cassin, 1968).[1]

However, something similar also occurred for biological anthropology, human genetics, and evolutionary thinking more in general[2]. But here things were somehow more complex.

Although it was the Nazi atrocity that inspired biology's shift, it was not enough to repudiate Nazism itself. It would have been simple for anyone, eugenicists included, to condemn murder, but even a superficial glance confirms a deep connection between biomedicine, human genetics, and Nazi ideology. Therefore post-war biologists had to do more than condemn. They also had to distance themselves from some of the conceptual views that made that science possible.

Germany under the Nazis was a racial state (Burleigh and Wippermann, 1993) – a political organization in which the health of the race assumed a central value (Weindling, 1989; Müller-Hill, 1988; Deichmann and Müller-Hill, 1994). Nazism understood itself as a "biocracy," whose supreme goal was to cure and purify the national body from racial pollution by all possible means, including mass murder (Lifton, 1986; Bonah et al., 2006). Hitler was seen as the national surgeon, a characterization so profound that eugenic sterilization was popularly called *Hitlerschnitt* – "Hitler's cut" (Bock, 1983).

In Germany, the racial hygiene movement (*Rassenhygiene,* German's equivalent of "eugenics") was founded in 1905 by Alfred Ploetz. Eugenics and sterilization measures largely predate Nazism. However, the rise of Nazism saw a much deeper incorporation of biological language into state ideology (Proctor, 1988; Proctor, 1999 especially chapter 3; Weindling, 1989; Weiss, 1990, 2010). Nazism monopolized the discourse of racial hygiene for its own goals (Weindling, 1989). For instance, the teaching of genealogy in schools was not unknown before the Third Reich, but after 1933, it was put to new and universal use to inculcate in students the ideological power of heredity and race (Weiss, 2010).

The regime self-consciously understood Nazism as "applied biology." The definition was coined by Fritz Lenz and subsequently made famous by Rudolf Hess (see Proctor, 1988: 62; Lerner, 1992; Kühl, 1994). Slogans such as "biology and genetics are the roots from which the National Socialist world view has derived its knowledge, and from which it continues to derive new strength" and "National Socialism without a scientific knowledge of genetics is like a house without an important part of its foundation" were essential to Nazi propaganda (see respectively, Proctor, 1988: 10 and 84). Nazism, it was said, wanted to "liberate the genotype of the German people" (Duello, 2010). In his 1943 article "Heredity as Destiny," Eugen Fischer – the longtime director of the Kaiser Wilhelm Institute of Anthropology, Human Heredity, and Eugenics and a highly regarded geneticist, even in the English-speaking world – publicly expressed his satisfaction with the flourishing of genetics research under the Nazis, explaining how the immediate function of this newly accumulated knowledge was "to serve the policy of the state" (quoted in Weiss, 2010: 119).

Human heredity was considered the foundation of the Reich from a practical as well as an ideological perspective. On the same day the Nazis outlawed all other political parties in 1933, the government passed a Law for the Prevention of Genetically Diseased Offspring (Bock, 1983; Proctor, 1988; Weiss, 2010). Its original title was the Law Against the Propagation of "Lives Unworthy of Life" and sanctioned the use of force "against those who did not submit freely" (Bock, 1983). "Strongly influenced by American models," (Kühl, 1994) it was only the first in a long series of racial hygienic measures. The following year, nearly two hundred genetic health courts (*Erbgesundheitsgerichte*) were established to evaluate cases for sterilization (Bock, 1983), of which there would eventually number almost 400,000 (almost eight times the effects of American sterilization laws). The periodical *Der Erbarzt* (*The Doctor of Heredity*) was founded in 1934 with the goal of making, in the words of the biologist Otmar Freiherr von Verschuer, "every doctor a genetic doctor." That same year the prestigious journal *Zeitschrift für Morphologie und Anthropologie* dedicated a special issue to Fischer's work, with the following introduction:

> We stand upon the threshold of a new era. For the first time in world history, the Führer Adolf Hitler is putting into practice the insights

> about the biological foundations of the development of peoples – race, heredity, selection. It is no coincidence that Germany is the locus of this event: German science provides the tools for the politician. (quoted in Marks 2010)

Among the essays that followed, Marks notes, "were contributions from two Americans, Raymond Pearl and Charles Davenport" (Marks, 2008: 10).

Americans had a critical role in inspiring Nazi policies (Kühl, 1994). German racial hygienists' admiration for American achievements in sterilization dates back to the beginning of the twentieth century, and in the Weimar Republic translations of mainline American eugenicists were common (Proctor, 1988; Weindling, 1989; Kühl, 1994). Historian Reinhold Muller wrote in 1932, "Racial hygiene in Germany remained until 1926 a purely academic and scientific movement. It was the Americans who busied themselves earnestly about the subject" (quoted in Proctor, 1988: 98).

Figures such as Harry Laughlin, who was awarded an honorary doctorate by the University of Heidelberg in 1936; Paul Popenoe, who regularly published in German racial hygiene journals; and Clarence G. Campbell, who represented American eugenics at the 1935 Berlin conference on Population Sciences, linked the American eugenics and German racial hygiene movements since the 1910s (see Kühl, 1994, 2013). The Austrian vice-consulate in California, Geza von Hoffmann, propagandized American eugenics successes in Germany and Austria well before the Nazis took power. Laughlin and Davenport's Eugenics Record Office maintained close contact with Ernst Rüdin, the prominent psychiatrist and eugenicist, and supported the hard-line German position on compulsory sterilization (Weindling, 1989: 504). In *Mein Kampf* (1924) Hitler himself applauded the American Immigration Restriction Act. Lenz, in his 1931 *Human Selection,* approvingly quoted Laughlin's forecast that 15 million individuals of inferior racial stock had to be sterilized by 1980 (Proctor, 1988). And public figures in American eugenics expressed admiration for the achievement of Nazi sterilization policy. In 1934 the secretary of the American Eugenics Society claimed:

> Many farsighted men and women in both England and America have long been working earnestly toward something very like what Hitler has now made compulsory. (quoted in Kühl, 1994)

This proximity between Nazi policies and American eugenics became particularly visible during the Nuremberg trials. In the medical trial, Karl

Brandt, Hitler's escort surgeon and a leading figure in the euthanasia program, referred explicitly to Grant's *The Passing of the Great Race* and to the sterilization laws in the United States and Scandinavia as direct sources of inspiration for the Nazi eugenic program (Weindling, 2004). As Kühl writes (1994):

> In their defense, those accused referred to the acceptance of the scientific basis of their work outside Germany. This strategy was based on the claim that democratic states had provided a model for the Nazi race policy. Physicians accused of organizing the "euthanasia program" in Nazi Germany pointed to the United States to prove that elimination of "inferior elements" was not unique to Germany. The 1927 United States Supreme Court decision affirming the legitimacy of eugenic compulsory sterilization in the United States was used by a German doctor as an example of the precedents for Nazi racial hygiene.

The involvement of surgeons Alexis Carrel and Edwin Katzenellenbogen in building eugenics programs in Vichy France and Germany, respectively, was highly embarrassing to Allied prosecutors. Both had been trained in America.

The way forward, as far as biologists were concerned, was to demarcate between "reasonable" and humane eugenics and the merciless "anti-science" of the Nazi program (Weindling, 2004; Proctor, 1999). C.P. Blacker's 1952 article in *The Eugenic Review* (Weindling, 2004: 324) exemplified this move, which became part of the official historiography of the eugenic movement. Blacker, in his official vest of secretary of the Eugenics Society, protested against the unfortunate connection "established in the minds of many people, including the War Crimes Commission" between eugenics and the Nazi "racialist practices." Galton's position, he objected, "which is now printed on the cover of our [Eugenics] Review, was "that eugenics was, in essence, a merciful creed, to be held by men endowed with pity and kindly feelings. It is therefore both unjust and deplorable that the word eugenics should be connected with Nazi racialist practices" (1952a: 9). The article was tellingly titled "'Eugenics' experiment conducted by the Nazis on Human Subjects", where the word "Eugenics" was obviously put in inverted commas. Galton's quotation (1908) appeared for the first time on the cover of the Eugenic Review in April 1947; it stated that "Man is gifted with pity and other kindly feelings; he has also the power of preventing many kinds of suffering. I conceive it to fall well within his province to replace Natural Selection by other processes that are more merciful and not less effective. This is precisely the aim of Eugenics." When this new cover appeared, the Nuremberg Trials were just underway.

According to Kühl (1994):

> After World War II, members of the American Eugenics Society sought to distance themselves from their former support for Nazi race policies ... Maurice A. Bigelow's "Brief History of the American Eugenics Society," published in *Eugenic News* in 1946, did not mention the Society's former support for Nazi attempts at race improvement. Neither did Frederick Osborn's "History of the American Eugenics Society" published in 1974. Reform eugenicists' artificial distinction between favorable parts of Nazi race policy and parts that needed to be condemned or concealed influenced their self-perception after 1945. The fact that they had criticized elements of Nazism allowed them conveniently to "forget" their prior support for Nazi eugenic racism.

Making of a new moral economy: genetics is democracy

While some biologists resorted to denial and hair-splitting, others crafted a new discourse that would leave prewar eugenics behind – not just assert compatibility with liberal democracy while maintaining the illiberalism of eugenics but develop a scientific reasoning for that compatibility.

In 1945 David C. Rife, a zoologist and geneticist, published *The Dice of Destiny*, whose last chapter was dedicated to "Genes and Democracy," (1945) a combination then without precedent, as far as I am aware. Montagu criticized Rife for his biologism and Dobzhansky took him to task for his lack of understanding of plasticity. Moreover, Rife still saw a conflict between genetic variation and democracy (Beatty, 1994). Nonetheless, the book was one of the earliest signs of an "awakening" of scientists after the "calamity of World War II" (Dobzhansky, 1945). In particular, the chapter "Not Two Alike" was among the first in a long series of attempts to align genetics with democracy, insisting on the scientific *cum* political value of individual differences, a theme that would become central to Dobzhansk's worldview.

Yale Review published a similar article that year. In "*The Biological Basis of Democracy*," Edmund Sinnott – botanist and future coauthor, with Dunn and Dobzhansky, of the 1950 textbook *The Principles of Genetics* – observed that "fascist totalitarianism or proletarian dictatorship captured the imaginations of many eager souls." To prevent this, he thought, any plan for a new society must find "sponsorship" not in emotions or abstract values but in scientific facts. He was therefore comforted by the knowledge that

> the free way of life, the true democracy...is so in harmony with the biological basis of the life of man, that it is better than the systems [fascists or Soviets] support.... It is rather in the fundamental character of protoplasmic structure and activity that we can find the basis for those essential aspects of democracy – freedom, progress, and the worth of individuals.

Democracy and its desire for freedom, in other words, were written into our very chromosomes, "in protoplasmic mechanisms which insure that each of us is different from his fellows," Sinnott wrote. Mendelian mechanisms insured "a continual re-shuffling of these genic differences and thus a much increased variety of types":

> The plain fact is and let us thank Heavens for it that the basis of our diversity is sunk deep in the constitution of living stuff itself, safe from totalitarian attempts to regiment us into an army of standardized robots who would march off the assembly lines of indoctrination as monotonously alike as a string of jeeps. (Sinnott, 1945/1946; see also Beatty, 1994)

The irreducible political role of genetics was to ensure diversity at each generation.

Dunn and Dobzhansky's 1946 *Heredity, Race, and Society* (see 1952 edition) introduced a wider public to the conceptual shifts of post-war biology. The book's key notion was that biology consisted in differences and variations, not hierarchies. Genetics, in particular, was a celebration of individuality:

> The chance that any two human beings, now living or having lived, have identical sets of genes is practically zero, identical twins always excepted. The hereditary endowment which each of us has is strictly his own, not present in anybody else, unprecedented in the past, and almost certainly not repeatable in the future. A biologist must assert the absolute uniqueness of every human individual. This same assertion, translated into metaphysical and political terms is fundamental for both ethics and democracy. (1946: 45–46)

Genetic diversity, far from "undermining democracy," was now seen "as a rationale for democratic equality," an argument that became mainstream in post-war "politico-genetics" (Dobzhansky, 1945).

As Dobzhansky repeated almost twenty years later:

> Denial of equality of opportunity stultifies the genetic diversity with which mankind became equipped in the course of its evolutionary development. Inequality conceals and stifles some people's abilities and disguises the lack of abilities in others. Conversely equality permits ... an optimal utilization of the wealth of the gene pool of the human species. (1962: 285)

It is against this background, featuring many approaches to race after the war, that the United Nations attempted to codify a new notion of race. Unsurprisingly, with such a range of views – from revisionist eugenics to genetic democracy and individual – vying for supremacy, the attempt to settle on a single concept of race proved controversial.

UNESCO and the contradictions of post-war race theory

The UNESCO statements on race, and the debates surrounding them, illustrate the novelties and tensions of post-war biology. They were the most visible application of "the spirit of the Modern Synthesis" in the public sphere (Gayon, 2003) and relied on a central ideology that might be called "racial liberalism" (Proctor, 2003). Racial liberalism had a remarkable capacity to "carry multiple meanings so that different audiences could discover congenial interpretations" (Selcer, 2012). It also embodied a significant dilemma of post-war biology, that "between objective science and [the] correct moral position" (Marks).

The same venue where this debate took place is very telling. The UNESCO was established with the idea to embody a "common culture of international understanding following the defeat of Nazism" (Blue, 2001). Founded in November 1945, UNESCO represented a sort of "secular soteriology" (doctrine of salvation), based on ideas of moral ransom from Nazi racial doctrines (Stoczkowski, 2009). Especially in its first phase (ibid.), UNESCO stressed ideas of universalism and unity of humankind, putting "humanity as a whole at the center": "What is not good for Humanity as a whole cannot be good for any nation, race or individual" was one of the leitmotiv of Jaime Torres Bodet the Mexican diplomat who replaced Huxley as UNESCO secretary in 1948. Many leading figures in the UNESCO were profoundly forged by war devastations, one case being Alfred Métraux (1902–1963) a Swiss-American anthropologist who, as a member of the U.S. Bombing Survey, was well aware of the deep destructions of war in Europe. Looking at these devastations reinforced "belief in the necessity for European unity and for the need of a firm basis for international, inter-cultural, and inter-racial understanding. His early view of war devastated Europe was important in his decision in 1946 to take a post on the secretariat of the United Nations" (Wagley, 1964).

The UNESCO statements on race

The debates over the statements are a play in four acts, but the first two, the 1950 statement and its 1951 rejoinder, are the more intellectually significant. The 1950 statement largely represented the social-science view on race that emerged from the post-Nazi scenario. Its architect was Ashley Montagu, a well-known Boasian anthropologist who had published a few years before his popular *Man's Most Dangerous Myth: The Fallacy of Race* (1942). Montagu headed a committee of experts drawn from Brazil, France, India, Mexico, New Zealand, the United Kingdom, and the United States. Anthropologists such as Claude Levi-Strauss and Juan Comas worked alongside the likes of Morris Ginsberg, editor of the leading British journal *Sociological Review*. The committee's task was to write "a statement defining the notion of race and setting out in clear and simple terms the present state of our knowledge on the oft-disputed problem of race equality" (Métraux, 1951; see also 1950a and b). The final text was revised by Montagu following a round of criticisms by geneticists and biologists including Dunn, Dobzhansky, Huxley, and Muller.

The *UNESCO Courier*, official magazine of the organization, celebrated and summarized the 1950 statement under the title "Fallacies of Racism Exposed: UNESCO publishes declaration by world scientists":

> False myths and superstitions about race contributed directly to the war, and to the murder of peoples which became known as genocide – but victims of the war were of all colours and of all "races." Despite the universality of this agony and destruction, the myths and superstitions still survive – and still threaten the whole of mankind. The need for a sound unchallengeable statement of the facts, to counter this continuing threat, is a matter of urgency...
>
> UNESCO offers this declaration as a weapon – and a practical weapon – to all men and women of goodwill who are engaged in the good fight for human brotherhood.

Race, according to the statement, was not a biological phenomenon but a "social myth." All ethnic groups had the same mental capacities. *Pace* right-wing Mendelians from Davenport to Lenz, race mixing was not biologically wrong.

Montagu's powerful worldview, with its tensions and contradictions, shaped these key points. Lines such as "national, religious, geographic, linguistic and cultural groups do not necessarily coincide with racial groups" and "the cultural traits of such groups have no demonstrated genetic connection with racial traits" asserted that race was a purely technical notion with no cultural implications. The Kroeberian move of disjoining via hard heredity the sociocultural from the biological was further radicalized. This demystification of race – a process Huxley started in 1936 – prompted Montagu to simply get rid of the term altogether and instead use "ethnic group." This phrase, lacking the emotional and political charge of "race," would be more appropriate to describe a merely technical phenomenon. Though the 1950 statement understood race as a social myth, it also recognized race as a "biological fact" (differences in gene frequencies in certain populations), but for "all practical purposes" the first meaning dominated.

Dunn and Dobzhansky did not go quite so far as to dispense with the word "race." Instead they redescribed race in much the same demystified and de-socialized terms as Montagu described ethnic groups: "populations which differ in the frequencies of some gene or genes" (Dunn & Dobzhansky, 1952: 118). This prompted Montagu to ask why anyone should use the "antiquated, mystical conception of race" to "describe populations in terms of their gene frequency differences," (1962b) to which Dunn and Dobzhansky replied that any term, including "ethnic group," might become loaded with intense, possibly destructive, social and moral values.

The divergence was superficial. Both sides shared the view that genetics had dislodged race from the Lamarckian biocultural kingdom, with its mystical union of habits and biology, and neatly separated it from the mental (Kroeber, see intermezzo I; Stocking, 1968). Genetics and the modern synthesis had "changed the meanings of basic terms" (Farber, 2010) and only because of this, race (or ethnic group) could be reduced to the variations of gene frequencies in different populations.

Both Montagu and Dobzhansky shared a post-Weismannian/Kroeberian framework in which it was possible to demarcate clearly the biological from the social. In the UNESCO magazine feature on the statement, the Swiss anthropologist Alfred Métraux endorsed a similar move:

> Race prejudice thrives on the inability of most people to make a clear distinction between facts pertaining to civilization and culture on the one hand and biological facts on the other. Men are distinguished by their respective cultures... the real differences between human societies are not due to biological heredity but to cultural environment.

In this post-war vision, the biological had not disappeared, however. It remained stably as a source of likeness and universality, while culture alone was given the power to diversify humankind. As the first statement said, quoting Confucius, "Men's nature are alike; it is their habits that carry them far apart." This biological likeness was as important as cultural differences, indicating biologism's subtle persistence.

This biologism is revealed throughout the statement, not least in the claim that "biological studies lend support to the ethic of universal brotherhood," which reveals the profound contradictions between two of Montagu's goals. On the one hand, he aims to disentangle mental characteristics from innate or biological factors (1972: 82–83). On the other, he aims to biologize virtuous mental and behavioral characteristics, such as the "drive to cooperation," now thought of as universal (see also Weidman, 2012). Whereas the statement nullified biology as a local, that is to say racial, experience, at the same time it reintroduced biology as a solid basis for universal positive traits and a warning to respect the profound dictates of human nature: "For man is born with drives toward cooperation, and unless these drives are satisfied, men and nations alike fall ill" (This strong concept of biological and universal human nature gained importance after the 1970s as we shall soon see).

The scientific community was far from persuaded by the statement. The epicenter of opposition was probably the journal *Man*, based in London, which primarily published works related to physical anthropology. Shortly after the statement's publication, the journal lamented:

> the manner in which the whole vast field of racial studies, physical as well as cultural, was thrown open to discussion by a small group of philosophers, historians, sociologists and others, only two of whom had any pretensions to competence in physical anthropology. (1950)

The statement's conclusions about race seemed unjustified "in the present state of our knowledge," physical anthropologist William Fagg wrote in a letter to the London *Times* (quoted in Selcer, 2012).

Amid the controversy, Dunn was appointed to coordinate a second statement in 1951. This time, biologists were more directly involved, and Montagu's role was much reduced. In addition to Dunn, Haldane, Huxley, Dobzhansky, and other biologists signed the Statement on the Nature of Race and Race Differences.

After a reassuring confirmation that all men belonged to the same species, "even though there is some dispute as to when and how different human groups diverged from this common stock," the text reintroduced the centrality of race. As Dunn commented:

> The physical anthropologists and the man in the street both know that races exist; the former, from the scientifically recognizable and measurable congeries of traits which he uses in classifying the varieties of man; the latter from the immediate evidence of his senses when he sees an African, an European, an Asiatic and an American Indian together. (UNESCO, 1969: 37)

This language represented a tactical move intended to please the wider public and physical anthropologists, as well as a complicated compromise among very different disciplinary communities. Certainly, the statement showed a high degree of diplomatic finesse:

> In its anthropological sense, the word "race" should be reserved for groups of mankind possessing well-developed and primarily heritable physical differences from other groups. Many populations can be so classified but, because of the complexity of human history, there are also many populations which cannot easily be fitted into a racial classification.

As Gayon (2003) has noted, the second statement was more cautious on race, eliminating emphatic claims made by the first – e.g., that "biological studies lend support to the ethic of universal brotherhood." To accommodate the requests of many scientists who complained that Montagu confused *is* and *ought* – scientific findings and normative values – the second statement divorced data and social values. "Equality of opportunity and equality in law in no way depend, as ethical principles, upon the assertion that human beings are in fact equal in endowment," the second statement read.

The third statement, published in Moscow, 1964, and the fourth, in Paris, 1967 are less relevant here, mainly because they express not turmoil but hegemony. Selcer correctly observes that, by the time of the third and fourth statements, the anti-racist hegemony had largely consolidated. The 1964 statement, "Proposals on the biological aspects of race," was "relatively uncontroversial," and attempts to rehash the 1950–1951 dispute failed (Selcer, 2012). It is interesting, however, that the third statement was flexible enough to attract even the signature of Carleton Coon, one of the few racialist thinkers of the period. The fourth statement ("Statement on race and racial prejudice") aimed to analyze racism and racist doctrines rather than the biological meaning of race.

Things had moved rapidly in the years since the first statement. How in only a decade and a half did this novel landscape emerge? In fact, there were four major points of conceptual transition. As we shall see in the next chapter, these demarcated (in the sense of boundary-work, Gieryn, 1991) good, fully democratic science from potentially totalitarian, anti-democratic science. These were scientific shifts, but they were molded by societal pressures, and, at the same time, shaped the global political agenda.

6
Four Pillars of Democratic Biology

The repositioning of biology within a liberal-democratic framework required more than repudiation of eugenics and scientific racism. Important conceptual transitions and a new science consolidated the change in political rhetoric.

In this chapter, I focus on four pillars of the post-1945 scenario. First, there was the classical-balance controversy, which pitted Dobzhansky against Muller and had considerable implications for eugenics and understandings of human heredity. Second, evolutionary thought was (allegedly) reconstructed in terms of population thinking and visibly disassociated from previous typological views of race. Third, there emerged the idea of a human culture either autonomous from biology or else representing a stage of evolution beyond that attainable from other animals. Either way, biologists, in a new spirit of cooperation with anthropologists, asserted human uniqueness. Fourth, organicism, the notion that groups constitute whole organisms unto themselves, eroded in the name of a new individualistic view of biological processes – something that was profoundly resonant with liberal-democratic values. At these points of transition, a new post-war democratic science found a consistent script by which it could differentiate itself from the anti-democratic biology of the past.

To test the tenacity of this new democratic framework, at the end of the chapter I investigate the return to some theoretical assumptions of interwar eugenics in sociobiology and evolutionary psychology. As we will see, even in that case, the global democratic framework prevailed, rendering docile and in the end anti-eugenic the conservatism of sociobiology and evolutionary psychology.

Democratic repositioning I: the classical-balance controversy

The so-called classical-balance controversy (covered in Lewontin, 1974; Beatty, 1987; Paul, 1987) refers to a technical debate in population genetics, but one with profound policy and moral implications. "Embedded ideological assumptions" played a crucial part in this dispute (Lewontin, 1974: 157) between two of the giants of twentieth-century genetics, Muller and Dobzhansky. Muller, a Nobel Prize on his mantle, was back in America after his tragic Soviet experience, first teaching at Amherst and then at Indiana University. The post-war Muller was still a eugenicist but of a post-revolutionary sort. He maintained faith in the control of human evolution, but without recourse to socialism. He withdrew in 1948 from the Soviet Academy of Sciences, at the peak of the Lysenko controversy (the letter of resignation and the Soviet official reply are worth reading for the intense politicization of science in both arguments, see Zirkle, 1949).

Although their paths repeatedly crossed and their alignment with classical genetics was similar, Dobzhansky was, in many respects, the antithesis of Muller. An ethnic Ukrainian born in the Russian Empire, Dobzhansky trained under Yuri Filipchenko in Petrograd and emigrated to the United States in 1927 to work with Thomas Hunt Morgan. Dobzhansky never returned to the Soviet Union, where he was labeled an enemy of the people. As author of one of the key texts of twentieth-century evolution, *Genetics and the Origin of Species* (1937), Dobzhansky was not only one of the most influential twentieth-century evolutionists and a key figure in the construction of the modern synthesis, but the driving force behind biology's embrace of the liberal-democratic order (for biographical treatments of Dobzhansky, see Provine, 1981; Levine 1995; Adams, 1994; Kohler, 1994; Smocovitis, 1996; Ayala, 1976, 1985; for rhetoric Ceccarelli, 2001; for his ideological aspect Krimbas, 1994; Beatty, 1994; Lewontin, 1974).

In 1955, Dobzhansky coined the term "classical-balance" in an attempt to distinguish his position on population genetics from Muller's. Dobzhansky identified the "classical" side, personified by Muller, as the older, traditional view. In opposition to this, the "balanced" view, represented by Dobzhansky himself, was not only modern, but also, as the term implies, stable, not radical. Muller contested the terminology, but his efforts failed. As a rhetorical strategy, Dobzhansky's was a perfect choice: the term "balanced" conveyed to readers the sense of the repositioning of biology within a liberal-democratic framework, and away from the extremism of the eugenic phase.

Muller's "classical" view was in fact a radical one (1950). It could easily have been called the eugenical view of population genetics, or the "gloomy view," since it was a repackaging of older concerns about genetic decay. Its most sophisticated form is a text significantly titled "Our Load of Mutations" (1950). Here, the eternal eugenicist, five years after Hiroshima (nuclear radiation was a key preoccupation for Muller), was concerned once again with the dysgenic effects of civilization and the need to counter the biological decline of humankind through intervention in the genetic composition of the human population. Muller's point was that it was not enough to intervene "euthenically," through medicine or other environmental avenues. This was prewar eugenics, cleverly rebranded in such a way so that the word eugenics never appeared in the text (apart from a reference to a eugenics conference in the early 1920s) and was instead replaced by the more neutral "artificial selection," along with a new terminology of "genetic loads" and "mutational loads." According to Diane Paul, in a letter to American Eugenics Society President Frederick Osborn,

> Muller noted that he had purposely chosen an apparently neutral title for his 1950 essay: an argument explicitly tagged with the eugenics label would have been dismissed in advance by many whom it did, in fact, influence. Words are not so important. The crucial thing, he argued, was to induce people to *think* eugenically. (1987: 328)

In the article, Muller opposed contemporary medical opinion, which held that mutation was a negligible cause of disease. There were plenty of harmful mutations in human populations, he said. However, the detrimental effects of these mutations were often hidden because:

> the mutant genes of a given locus usually produce, in any single individual, but a very small effect when heterozygous, but accumulate until they reach a reciprocally high frequency in the population, and so do as much total damage as if they were completely lethal. It is calculated that the average individual is probably heterozygous for at least 8 genes, and possibly for scores, each of which produces a significant but usually slight detrimental effect on him. (1950: 170)

Repeating a typical eugenic anxiety, Muller claimed:

> The improvements in living conditions, medicine, etc. under our modern civilization must result in a saving for reproduction, at present, of a large proportion of those who under the earlier conditions would have been genetically proscribed.

In other words, contemporary social organization was giving free reign to harmful mutations that were once checked by the death of the unfit. Mankind harbored, "more or less unconsciously," a ticking genetic time bomb. Few were aware of the dangers, Muller thought, because, courtesy of the evolutionary checks of the past, the danger didn't seem imminent. Muller's contemporaries were becoming a "debtor generation" who, "by instituting for their own immediate benefit ameliorative procedures which delay the attainment of equilibrium and raise the equilibrium level of mutant gene frequency," transfer "to their descendants a price of detriment which the latter must eventually pay in full." It would be "disaster for mankind."

In a chapter of his *Mankind Evolving* (1962), titled "Muller's Bravest New World," Dobzhansky wrote that prophets of doom invariably include prescriptions for its avoidance. In Muller's case, this meant massive political intervention at the genetic level; that is, eugenics:

> The only means by which the effects of the genetic load can be lightened permanently and securely is by the coupling of ameliorative techniques, such as medicine, with a rationally directed guidance of reproduction. (Muller, 1950)

In spite of the "social obstacles" preventing it, this was the most straightforward "escape" Muller offered, a policy prescription that could only be achieved through "a deep-seated change in mores." The alternatives were "in the long run, as effective as trying to push back the flowing waters of a river with one's bare hands."

Dobzhansky didn't share Muller's concern that "the genetic loads which human populations carry are unconditionally deleterious" (Dobzhansky, 1962: 343). Since at least the late 1930s (Dobzhansky, 1937), he had pioneered a different view of population genetics, which argued that "genetic variation and polymorphism" were "at the very center of the study of evolutionary dynamics" (Lewontin, 1997: 352).

This alternative school was now consolidating and bringing to the floor a view that was less gloomy and more optimistic, or "balanced," to again use Dobzhansky's term. It had profound anti-eugenic implications. From the point of view of the new population genetics, things were more in harmony than Muller had feared. The new school hinted that a genetic laissez faire, whereby leaving almost any genetic load unchecked, would be the best policy for mankind.

By the time Muller was rebranding eugenics in terms of mutational loads, Dobzhansky had been critical of "eugenical Jeremiahs" for years.

They keep "constantly before our eyes the nightmare of human populations accumulating recessive genes that produce pathological effects when homozygous," he wrote in 1937 . These "prophets of doom" could not see the evolutionary advantage of "the accumulation of germinal changes in the population genotypes." Only this vast genetic potential, a "supply of hereditary variation," could guarantee better adaptability to mutated environmental conditions – in a nutshell, evolutionary plasticity (ibid.).

But in 1937 Dobzhansky stopped short of drawing the full human implications of his view of population genetics. He eventually did so in 1955. "At a risk of oversimplification," he presented the two "working hypotheses" of the classical and the balanced view. According to the former:

> Evolutionary changes consist in the main in gradual substitution and eventual fixation of the more favorable, in place of the less favorable, gene alleles and chromosomal structures. Superior alleles are established by natural selection, and supplant inferior ones. (1955: 3)

This view implied that homozygosis (two chromosomes of a pair with the same gene arrangement) was the normal state, and heterozygosity (different gene arrangements on the two chromosomes) the abnormal one, mostly created by deleterious mutations and therefore to be corrected. The classical perspective buckled under the weight of normative ideas about optimality. The balanced view, by contrast, saw in heterozygosity an evolutionary advantage: evolution tended to prefer heterozygosity to homozygosis, "a genetic good mixer" over "a genetic rugged individualist" (1955:3; see also Lerner, 1954; Beatty, 1994).

Dobzhansky introduced a distinction between "two different loads of mutations that Mendelian populations have to carry" in order to counter Muller's view of genetic loads. There were indeed real burdens produced by genetic mutations, but these were much smaller factors than Muller thought. The bulk of these mutations were neutral or even beneficial for human populations, Dobzhansky claimed. He labeled Muller's genetic load "mutational": "recessive genetic variants deleterious to their carriers in most environments in which the population lives." But this was a secondary factor in evolution. The other category, the "balanced" load, referred to variations that were harmful in one genetic environment or under specific environmental circumstances but favorable in other settings (1955: 5–6). One textbook case of heterozygotic advantage is the gene for sickle cell anemia, which, in the heterozygous condition, offers the carrier protection against malaria.

The scientific-cum-political implications of Dobzhansky's work were obvious: "The production of deleterious mutants is the price which a living species pays for retaining the evolutionary advantages of genetic plasticity" (Dobzhansky and Wallace, 1954). Since many loads were balanced rather than mutational, and since genetic plasticity was advantageous, planned exclusion from the gene pool was never justified:

> These genes may be maintained in human populations in balanced states, either because of being advantageous in heterozygotes or because of the action of diversifying selection.

Only in extreme instances, when human suffering and the costs to society could not be overlooked, did Dobzhansky propose educating and informing affected individuals in the hope that they might freely decide not to have children. In cases of individuals mentally incompetent to reach such a decision, he considered obligatory sterilization (1962: 333, 1965).

Dobzhansky's view displaced the notion that certain human genotypes were normal or optimal (Medawar, 1960; Beatty, 1987). This implied a further shift in the understanding of fitness, now reduced from a value-loaded concept to a merely technical notion of "reproductive success" (Paul, 1988). If fitness retained ideological meaning, it was consonant with pluralism of values favored by liberal democracy. As Dobzhansky wrote in 1970, the classical model "assumes that there exists a 'normal type' of each species or population, carrying the 'normal genes,'" whereas the balance model

> acknowledges genetic diversity as a fundamental phenomenon of nature. The gene pool of a population is envisaged as an array of alleles at many, perhaps at most, gene loci. None of these alleles may be the universally "normal" ones; the fitness conferred by many of these gene alleles on their carriers depends on what other alleles at the same and at other loci are present in the genotype and, of course, on the environment in which the carriers develop and live. There is no "normal type," only an adaptive norm composed of an array of genotypes. (1970: 198)

The controversy, thus, was not merely technical but more profoundly ideological. As Richard Lewontin puts it, the disagreement reflected

> a divergence between those who, on the one hand, see the dynamical processes in populations as essentially conservative, purifying and

> protecting an adapted and rational status quo from the nonadaptative, corrupting, and irrational forces of random mutation, and those, on the other, for whom nature is process, and every existing order is unstable in the long run. (1974: 57)

Muller inverted this perspective. He thought of Dobzhansky as a conservative and attacked the balanced view as a mystical old notion favoring genetic mediocrity and decay. He disliked its laissez faire attitude, something that was so distant from his social engineering mentality. For his part, Dobzhansky celebrated genetic variations as the basis for a fitter humanity: variation should not be decreased but enhanced (Beatty, 1987). "Do we really want to live in a world with millions of Einsteins, Pasteurs, and Lenins?" he asked polemically having once again Muller (who had used these expressions in his 1935 *Out of the Night*) as a target (1962). A world of clones, even of the best type, would not be a democratic world. Democracy, like genetics, required variations to work well. A first step toward the re-alignment of genetics and democracy was therefore accomplished.

Curiously, the dispute between Muller and Dobzhansky was never closed. Rather, as a new generation of evolutionary theorists emerged in the 1970s and '80s, the debate lost its audience. Whereas for both Muller and Dobzhansky the debate was inextricably scientific and political, the next generation saw a merely technical controversy, disconnected from broader social issues (Paul, 1987; see also Beatty, 1987).

Democratic repositioning II: constructing population thinking

The transition from typological to population thinking, advanced by Ernst Mayr, was a second critical moment of biology's post-war repositioning. Population thinking should not be confused with the biological concept of population (Hey, 2011), which refers to "actually or potentially interbreeding natural populations which are reproductively isolated from other such groups" (Mayr, 1942). Rather, according to population thinking, population is just a fictional set of varying individuals. This view exists in opposition to typological thinking, wherein population is itself a reified entity or type.

Mayr began to articulate population thinking in 1942, and in polemical terms. There was a clear hero: Darwin. Mayr believed he had displaced typological and essentialist views in favor of populational ones. There was a clear villain: Plato. Mayr believed his notion of *eidos* (visible form) implied typologist or essentialist views (Mayr used the two terms interchangeably, see 1982), which dominated Western thought for two millennia. Although there was a long transition period beginning in 1942 in which many of the "rudiments" of population thinking took shape in Mayr's thought (Chung, 2003), he made the distinction explicit only in 1959. In "Typological versus Population Thinking," Mayr argued that typological thinking offered

> a limited number of fixed, unchangeable "ideas" underlying the observed variability [in nature], with the *eidos* (idea) being the only thing that is fixed and real, while the observed variability has no more reality than the shadows of an object on a cave wall. (1959, reprinted in 1997)

According to Mayr, had Darwin believed this, theories of evolution and natural selection would never have been formulated. He had to have been a populational thinker (1982: 47), who recognized uniqueness of everything in the organic world:

> All organisms and organic phenomena are composed of unique features and can be described collectively only in statistical terms. Individuals, or any kind of organic entities, form populations of which we can determine the arithmetic mean and the statistics of variation. Averages are mere abstractions; only the individuals of which populations are composed have reality. (1959)

The two worldviews were incommensurable:

> The ultimate conclusions of the population thinker and of the typologist are precisely the opposite. For the typologist the type (*eidos*) is real and the variation an illusion, while for the populationist, the type (average) is an abstraction and only the variation is real. No two ways of looking at nature could be more different. (1959)

These two ways of thinking led to stark differences in interpretation of biological phenomena, such as race. For the typologist, "Every representative of a race conforms to the type and is separated from the representatives of any other race by a distinct gap." For the populationist, race

> is based on the simple fact that no two individuals are the same in sexually reproducing organisms and that consequently no two aggregates of individuals can be the same. If the average between

> two groups of individuals is sufficiently great to be recognizable on sight, we refer to such groups of individuals as different races. (1959: 28)

The political overtones of this move were clear, and the emphasis on varying individuals rather than fixed types soon became "an important component of the antiracist script" (Selcer, 2012).

Not only race, but also natural selection can be understood either in a typological or in a populationist way. The former approach uses value-laden terms such as "good," "bad," "useful," and "detrimental" to describe variations. The latter uses the term "superior traits" only to define environmentally specific advantages that contribute "to the gene pool of the next generation." There is no effort to create or maintain a superior type. Here we find resonances with the classical-balance controversy: for the typologist, variation is error or deviation from a true type; for the populationist, variation is a neutral fact, simply the means by which evolution works (O' Hara, 1997).

The transition from typology to population thinking, much like the transition from soft to hard heredity, was construed as a modernization of evolutionary thinking, the replacement of old-fashioned modes of thought with new ones. Just as in the case of the soft-versus-hard-heredity debate, Darwin – in spite of his often embarrassingly typological language, including "forms" and "varieties" rather than "populations" and "individuals," when talking of species – was hailed as the hero for this 'modernization' (Mayr, 1982: 268).

The politics and "existential" appeal of populationism

It is not difficult to see the "existential" (Hey, 2011) and political appeal that populationism had at the time. In the post-war period, the opposition between typology and populationism served "good" Western biological science's twin goals: mostly, to demarcate itself temporally from its dark eugenic past but also spatially from its totalitarian-Lysenkoist competitor in the Soviet Union. Like liberal democracy, population thinking celebrated the power of individual difference and uniqueness while diminishing the value of overarching structures reflected in the typological emphasis on homogeneity.

Mayr exploited this convergence strategically. He repeatedly made clear that the tragedy of eugenics was its typologization. In this way he traced a clear genealogical line between typology and the gas chambers, Plato and Hitler. It was an argument redolent of Karl Popper's political philosophy, which was becoming increasingly influential in the United

States and the Anglo world at the time (see Winsor, 2006 for the influence of Popper on Hull and of this latter on Mayr). As Mayr wrote several decades later:

> No political bias was at first attached to eugenics, and it was supported by the entire range of opinion from the far left to the far right. But this did not last long. Eugenics soon became a tool of racists and of reactionaries. Instead of being applied strictly to population thinking, it was interpreted typologically; soon, without the show of any evidence, whole races of mankind were designated as superior or inferior. In the long run it led to the horrors of Hitler's holocaust. (1982: 623)

Interestingly, Mayr does not reject eugenics as such, but rather its typological application. In a letter to Francis Crick, dated April 14, 1971, Mayr matched population thinking and his persisting eugenics faith:

> I have been favoring positive eugenics as far back as I can remember. As I get older, I find the objective as important as ever, but I appreciate also increasingly how difficult it is to achieve this goal, particularly in a democratic western society. Even if we could solve all the biological problems, and they are formidable, there still remains the problem of coping with the demand for "freedom of reproduction," a freedom which fortunately will have to be abolished anyhow if we are not [*sic*] drown in human bodies. The time will come, and perhaps sooner than we think, when parents will have to take out a license to produce a child. No one seems to question that it requires a license for such a harmless activity as driving a car, and yet such an important activity as influencing the gene pool of the next generation can be carried out unlicensed. A biologist will understand the logic of this argument, but how many non-biologists would? Obviously, then, we need massive education.

In more general texts, Mayr highlighted the tensions between democracy and evolutionary thought. In his opus magnum, *The Growth of Biological Thought* (1982), he explains that democracy and its Enlightenment roots were not shaped by biological thought but rather by a mixture of "physicalism and antifeudalism," which "has taken over in the western world to such an extent that even the slightest implied criticism (as in these lines) is usually rejected with complete intolerance."

There are, according to Mayr, some common traits between "democratic ideology and evolutionary thinking," which share "a high regard for the individual." But the individual means something different in each case. Anyone who believes in the "genetic uniqueness of every individual" cannot believe in the dictum, "All men are *created* equal." Rather, one would have to conclude, "No two individuals are *created* equal."

As evident from these quotations, among the fathers of the modern synthesis, Mayr was the least enthusiastic liberal democrat. Recall, for instance, Dobzhansky's attack on Carleton Coon's *The Origin of Races* – Mayr instead reviewed the book favorably and seemingly apolitically (Mayr, 1962; Jackson, 2001; Collopy, 2015). "I saw none of the implications which you seem to see," Mayr wrote polemically to Dobzhansky.

And yet Mayr was so adamant about the liberal-democratic populational-typological story that he was willing, essentially, to fabricate it. His assertions of Darwin as father of populationism and claims of essentialism dominating pre-Darwinian thought are dubious (Sober, 1980; Sloan, 1985; Greene, 1992; Amundson, 1998; Winsor, 2001, 2006; Stevens, 2002; Levit and Meister, 2006; Müller-Wille, 2007b; Hey, 2011; Powers, 2013; Witteveen, unpublished). Mary Winsor has argued that Mayr's story "simply grew with repetition." She highlights the "contrast between the enormity of its reputation and the flimsiness of its basis in historical evidence" (Winsor, 2006: 168, 150). The neat division of the history of biological thought between typological and populational thinkers does not stand analysis (ibid., see also Amundson, 1998). According to Elliott Sober (1980), other hypotheses coexisted with essentialism, which was itself open to dispute. Mayr even tried to impose his populational view on Darwin by underhanded means. In the index for a 1964 edition of *Origin of Species*, Mayr introduced a heading for "population thinking," which directed readers to two passages, neither of which justified inclusion. As Jody Hey puts it, "In short, Darwin did not provide us with any text that directly resembles Mayr's explanation of population thinking" (Hey, 2011: 259; see also Gayon, 1998: 117).

Darwin looked certainly at variation, but the value-laden philosophical framework Mayr employed, and the whole story of a transition from essentialism to populationism, was mostly a mid-twentieth century creation by Mayr himself. One might offer internalist explanations for this terminology; after all, biologists had new knowledge on which to draw – a hypothesis that certainly has its reasons. But, given the range of evolutionary thought and its public communication at the time, it is hard to deny that pressures external to biology itself helped to solidify the populationist worldview. The post-war world selected, via Mayr, the story most fitting its new underlying liberal-democratic ethos.

Democratic repositioning III: the humanism of the modern synthesis

Two highly visible features of biology's post-war repositioning concern the evolutionary understanding of what was then called "man".

The first was the growing integration of evolutionary biology, in particular the modern synthesis, with anthropology. Although this process began in the 1920s, it only consolidated after World War II. The architects of the modern synthesis – Simpson, Huxley, Mayr, and especially Dobzhansky – supported this integration. This befit a broader tendency to unify knowledge and phenomena: evolutionary mechanisms were one and the same no matter the level of selection, from gene to individual, population to species to culture (Smocovitis, 2012).

The second, one of the four pillars of the transition to liberal-democratic biology, implicated a philosophical gesture that practically all the architects of the modern synthesis eventually promoted, though with different nuances: the notion that human beings are not reducible to animal nature, because culture makes human life distinct within the animal kingdom. This commitment marked a profound break from the eugenic conflation of human mental traits with other bodily characteristics, and therefore with animality.

There was, however, a deep tension between the two projects, often obscured by labels such as "scientific humanism" and "evolutionary humanism" (Huxley, 1957; Smocovitis, 2009; Renwick, 2016). The first project emphasized a synthetic worldview capable of explaining all levels of evolution, from genes to cultural units. (E. O. Wilson's sociobiology was, in a sense, the heir of this effort.) The irreducibility project, on the other hand, emphasized the extraordinariness of man. In time, the second project was divorced from the first in the work of Lewontin, Gould (see respectively, Fracchia and Lewontin, 1999; Gould, 1997), and others who argued for dissimilarity between sociocultural and biological evolution. However, the strength of the post-war scenario lay, in some measure, in the capacity of biological thinkers to maintain both of these views at once.

Man's place in the new synthesis

According to Betty Smocovitis, "Anthropology, the discipline that dealt most immediately with human evolution, had been curiously removed from organizational and intellectual efforts to synthesize evolution in the 1930s and 1940s" (2012). It was only after the war, and in particular in the 1950s, that anthropology was integrated into the evolutionary synthesis, as illustrated by "watershed" meetings such as the 1950 Cold Spring Harbor symposium on "Origin and Evolution of Man" and the Darwin centennial at the University of Chicago in 1959 (Smocovitis, 1999, 2012; Little and Sussman, 2010; Little, 2012).

The Cold Spring Harbor meeting was organized by the geneticist Milislav Demerec with help from Dobzhansky and primatologist Sherwood Washburn. Simpson, Mayr, Coon, Dunn, and Montagu were among the participants. The attendees mostly concentrated on human origins and issues of racial differentiation. As the geneticist Curt Stern wrote in concluding remarks, "The political implications of statements or conclusions regarding the origin or evolution of man have been in our minds again and again" (Stern, 1950). It could not have been otherwise, given the symbolism of the location: the very place where Charles Davenport had organized the Eugenics Record Office (Little and Sussman, 2010). Sol Tax, editor of *Current Anthropology*, assembled the second meeting, which included panels dedicated to "The Evolution of the Mind" and "Social and Cultural Evolution" (1960).

A new anthropology in constant dialogue with evolutionary findings emerged from these interactions. Washburn in particular has been interpreted as a key figure in "remodeling" anthropology to meet the exigencies of the new "United Nations' post–World War II universal man" (Haraway, 1988). Donna Haraway has notoriously claimed that Washburn's "Man the Hunter" "embodied a socially positioned code for deciphering what it meant to be human ... after World War II," acting as "liberal democracy's substitute for socialism's version of natural human cooperation" (ibid.: 207).

Washburn's proposal for a "new physical anthropology" (1951) was a multidisciplinary attempt not simply to add "a little genetic terminology" to anthropology but to "change [anthropology's] ways of doing things to conform with the implications of modern evolutionary theory" (1951: 61). As Washburn realized, "Under the influence of modern genetic theory, the field is changing from the form it assumed in the latter part of the nineteenth century into a part of modern science" (ibid).

The new anthropology – integrated with the latest approaches in genetics and evolution, in particular population thinking – had vast impact (Boyd, 1950, 1953; see also Mayr, 1963: 646). Even a Boasian reconciliation with evolutionary thinking seemed possible. Cultural anthropologists Kroeber and White collaborated on the 1959 Darwin centennial and the volume produced by the conference considered man "*sub spaecie evolutionis*" (Tax, 1960). Kroeber's paper looked optimistically at the inclusion of "human history in total science" (ibid). White was more cautious and argued for the importance of the "extra-somatic" level (the former Kroeberian superorganic). Culture was only a human feature; it represented what he called the fourth stage of the evolution of mind. A well-defined notion of culture should remain at the center of any effort to understand human beings, White claimed:

> We must have a new science: a science of culture rather than a science of psychology if we are to understand the determinants of human behavior. (ibid)

Two concepts were once again in tension: humans as unique and humans as part of nature. There was no way around the tension, because it reflected two goals that were essential to the architects of the synthesis, who wished to make room for a science of human culture *and* to integrate anthropology within evolutionary biology. Dobzhansky sums up:

> Man has both a nature and a "history." Human evolution has two components, the biological or organic, and the cultural or superorganic. These components are neither mutually exclusive nor independent, but interrelated and interdependent. Human evolution cannot be understood as a purely biological process, nor can it be adequately described as a history of culture. It is the interaction of biology and culture. There exists a feedback between biological and cultural processes. (1962)

Dobzhansky championed the movement on behalf of the exceptionalism of human life and autonomous cultural processes, the third pillar of liberal-democratic biological thought. This movement favored obviously a reconciliation of biology with the Judeo-Christian root of liberal democracy – Dobzhansky was an Orthodox Christian often troubled by metaphysical questions.

Exceptionalism had its roots more in politics than in science. It was a response to eugenics' faith in the total commensurability between physical and mental traits and to a radical naturalism that would deny the uniqueness of the human mind. Eugenicists thought intelligence and musical talent could be bred like red pigeons. Promoting the specificity and autonomy of human cultural traits, "the biological uniqueness of man," or "man as an extraordinary creature" therefore was a gesture of profound political and epistemic significance (ibid. and 1963).

Although exceptionalism did not perfectly match anthropologists' disjuncture of the sociocultural from the organic (Smocovitis, 2012), they shared a common platform. Work coauthored by Dobzhansky and Montagu illustrates this. Published in *Science*, and titled "Natural Selection and the Mental Capacities of Mankind," the 1947 piece signaled evolutionary biologists' emerging post-war defense of autonomous cultural processes:

> Man is a unique product of evolution in that he, far more than any other creature, has escaped from the bondage of the physical and the biological into the multiform social environment. This remarkable development introduces a third dimension in addition to those of the external and internal environments – a dimension which many biologists, in considering the evolution of man, tend to neglect. The most important setting of human evolution is the human social environment. (Dobzhansky and Montagu, 1947: 587)

This crypto-dualist position, distinguishing the developmental sources of human mental and physical functions, was rare among biologists during the hegemony of eugenical thinking in the interwar period, either on the left or on the right. Recall that Mendelians (both left and right) believed in the dependence of mental traits on hereditary factors, and so did (the due epistemic changes made) Lamarckians. Kammerer even thought that education was an organic technique. Muller wanted to construct a country of Lenins and Einsteins via artificial insemination. Therefore the idea of "escaping from the bondage of the physical and the biological" sounded like a significant post-war novelty, or possibly a return to a pre-eugenic view of human uniqueness, as in the work of the anti-eugenicist co-father of evolution Alfred Russel Wallace.

Montagu and Dobzhansky's article was designed to please both sides, evolutionary biologists and anthropologists. Humanity was unique because of its malleability, the article claimed in keeping with an age-old humanistic tradition. From Pico della Mirandola to Vico, from Kant to Herder, plasticity was a key source behind anthropological thinking (Zammito, 2002; Meloni, 2011). However, to please the evolutionists, humanity was unique and plastic *because of its genes* and special biological adaptation:

> In general, two types of biological adaptation in evolution can be distinguished. One is genetic specialization and genetically controlled fixity of traits. The second consists in the ability to respond to a

> given range of environmental situations by evolving traits favorable in these particular situations; this presupposes genetically controlled plasticity of traits. (1947, 588)

This was a miracle of intellectual dexterity.

Dobzhansky made similar arguments elsewhere. The first chapter of *Mankind Evolving* (1962) was dedicated to "Biology and Culture in Human Evolution" and incorporated Kroeber's superorganicism; the repudiation of cultural evolutionism by Boas, Ruth Benedict; and the notion of culture as a distinct level of evolution. But Dobzhansky disliked extreme culturalism, the position denying any link between organic and superorganic. The following passage is typical of this unstable balance:

> In producing the genetic basis of culture, biological evolution has transcended itself – it has produced the superorganic. Yet the superorganic has not annulled the organic. (1962: 21)

Such acrobatics could also be found in Dobzhansky's 1956 *The Biological Basis of Human Freedom*, where he aimed at a precarious middle point rather than "explain human affairs entirely by biology" or "suppose that biology has no bearing on human affairs" (110). Both extremes were wrong, he said.

Its title notwithstanding, *The Biological Basis of Human Freedom* was remarkably open to the human transcendence of biology: "the most important agents which propel human history are maintained in that history itself, not in that stuff of which human genes are made," Dobzhansky wrote (119). Showing his religious roots and quoting Dostoyevsky, he explained that no matter the "the evolutionary bonum," human freedom involved "the capacity to rebel against it" (1956: 129; see also Dobzhansky, 1969). He then closed the passage with a typical contortionism: culture has "its own laws, which are not deducible from, although also not contrary to, biological laws" (1956: 134). And in his 1964 article "Evolution, Organic and Superorganic" (1964), he again oscillated between nature and its transcendence:

> Inescapably, man's nature is in part biological nature. But man is more than a DNA's way to make more of DNA of a particular kind.... Man receives and transmits... not one but two heredities, and is involved in two evolutions, the biological and the cultural.

Simpson had a similar attitude (see Laporte, 2000), which he rendered in similarly tortured language. In his classic *The Meaning of Evolution* (1949), he defined human social organization as "a result of organic evolution, but…something essentially different in kind." The Kroeberian lesson of a difference between organic and superorganic seemed at this point well-established amongst biologists:

> Organic evolution and societal evolution must, then, be not only constantly compared in their common aspects as evolution but also constantly contrasted in their differences as sharply distinct sorts of evolution, even though one is the product of the other and is, indeed, its continuation by other means. (290)

In a 1962 *Science* article, "The Biological Nature of Man," Simpson repeatedly assigned human social and cultural life uniqueness "in kind and in complexity." Human language, according to Simpson, was a token of exceptionality, a system of communication "absolutely distinct" from any other in the animal kingdom. No language instinct here. "Man has a niche and an ecology," Simpson wrote, but he "stands upright, builds and makes as never was built or wrought before, speaks and may speak truth or a lie, worships and may worship honestly or falsely." (1962: 478)

The dominance of the integrative position articulated by Simpson, Dobzhanksy, Montagu, and others is revealed in the discipline of their few remaining eugenicist contemporaries. As Huxley said, opening his 1962 lecture before the Eugenics Society, "Man, let me repeat, is not a biological but a psychosocial organism." Thirty years prior, the term "psychosocial" was, to say the least, rare amongst biologists. Huxley used it more than ten times in the address (Huxley, 1962). This was probably Huxley's "opportunistic" strategy (Weindling, 2012) and a mere matter of lip service, but it nonetheless indicates how widespread were the rhetorical effects of the modern synthesis' compromise with liberal democracy.

Democratic repositioning IV: individualism and the new political economy of nature

The final pillar girding liberal-democratic biology became more visible in debates the mid-1960s and at that point emerged partly as a challenge to the modern synthesis itself. But its beginnings came much earlier, with the shift in post-1945 biological thinking toward the individual as "a unique and unrepeatable realization in the field of quasi-infinite possible genetic combinations" (Ayala, 1985). Individualism was inherent in both Mayr's attack on typology and Dobzhansky's denial of a "normal" genotype that might be privileged by eugenics policy. Individualism provided a direct attack on the idea that evolution requires a conception of fitness at the level of the race or species – the collective. This collective ethos guided eugenics, which subordinated the individual to the "good of the whole community" (Esposito, 2011).

The debate between organicism and individualism had profound political resonances. Evelyn Fox Keller (1988) emphasizes that each pole implies a distinctive set of values: "autonomy, competition, simplicity; a theoretical privileging of chance and random interactions, and the interchangeability (that is, equality) of units" on one side, and "interdependence, cooperation, complexity; the theoretical privileging of purposive and functional dynamics, and often a hierarchical organization" on the other.

In the German context, organicism in biology was represented by the "Call to 'Wholeness,'" as Anne Harrington puts it (1996). Holistic science was mobilized against the mechanized and atomized world wrought by modern science and technology. The scientific eccentric, Jakob von Uexküll (1864–1944), was a key thinker in the application of holism to biology. His vitalistic metaphors and concept of *Umwelt* (the surroundings, or perceptual world where an organism lives) in particular, had a profound influence on German philosophy, Heidegger firstly. Ayelet Shavit summarized his view:

> Uexküll attacked both individual freedom and competition in the Weimar Republic, and equality in the new cooperative Soviet Union. Both Democratic and Bolshevist states represented the masses rather than hierarchically organized groups, and their lack of internal order drastically reduced individual and group utility. (2004)

Uexküll's dislike for individualism emerged powerfully in an exchange with the racist and radical right winger Houston Stewart Chamberlain, in which Uexküll explicitly linked individualism and destruction of the organism, cancer or revolution (quoted in Harrington, 1996: 58 see comment also in Shavit, 2004). Uexküll argued for "perfect harmony" between the *Volk* and the state, understood as a well-functioning organism aimed at limiting the destructive power of the individual, (ibid.: 60; Shavit, 2004) a line that became fashionable when the Nazis took power.

American biological thought concerning group-level phenomena was politically very different. University of Chicago ecologists Warder C. Allee and Alfred E. Emerson (Mitman, 1988, 1992) considered population a superorganism: a "collection of single creatures that together possess the functional organization implicit in the formal definition of organism"

(Emerson). From this they took a pacifist and internationalist lesson, very far from von Uexküll's nationalist, right-wing biology. However, the resonance that this organismic and communitarian economy of nature had after 1940, when the world conflict had detonated, was such to provoke growing discredit to it. Although it is sensible to be nuanced when defining the position of the modern synthesis on issues such as group selection (Borrello, 2010), unquestionably its ethos was individualistic and certainly not organicist nor communitarian. The political rationale behind this attitude is easy to see. Given the association of holism with the German radical right and the rise of the Soviet Union as a supposed non-individualist, cooperationist society, Anglo-American biologists supporting the modern synthesis felt much more comfortable in prioritizing the individual. They feared that the superorganism's "higher levels of integration could only be achieved by suppression of subordinate levels" (Mitman, 1988): the individual was at risk of being mystically dissolved into the higher organism, a good analogy for totalitarianism.

Even though the antecedents of this more individualist way of thinking can already be found in Fisher (1930), the construction of a true ideology of individualism, played against totalitarianism as a group-centered philosophy of biology, did not appear until the early 1940s. Only then did biologists begin to challenge "social theories, like those of Fascism and National Socialism, which exalt the state above the individual," to quote Huxley.

The best example of this new trend in moral and political philosophy is an astounding November 1940 lecture delivered by George Gaylord Simpson at the Paleontological Society of Washington, D.C. The United States was not yet involved in the conflict overseas, but Simpson, the scientist who integrated paleontology into the modern synthesis (1944), was already at war with Axis biology's organicism and totalitarianism, as well as American biologists who still believed in a possible coexistence of organicism and democracy. Simpson's conference is an unsurpassed example of politicization of biology, probably to be put on the same level of Muller's letter to Stalin (although, obviously, in radically divergent political directions).

The lecture and article derived from it, "The Role of the Individual in Evolution," (1941) opened with a neat position:

> Whatever happens in organic evolution, or indeed within the whole realm of the biological sciences, happens to an individual. Genetic mutations occur in individuals. Individuals struggle for existence and fail or succeed according to their equipment and circumstances. It is

> individuals that reproduce and that exercise such selection of a mate as may be possible to them. These facts are so evident that it may not seem worthwhile to state them, and similar statements so exhaust the basic aspects of evolutionary theory that it may seem impossible to say more about the role of the individual in evolution. Nevertheless such statements of the obvious are not needless, because the obvious is so often forgotten.

Simpson's stance was as much scientific as political. The distinction between group and individual had "implications of the greatest importance, extending even into the political sphere."

His definition of the two worldviews was strictly connected to the role each assigned to the individual, and he endorsed democracy in biological terms:

> The essence of democracy is belief in the importance and independence of the individual, and in the progress of society through the satisfactions of the individuals composing it. The essence of totalitarianism is belief in the unimportance of the individual and his subordination to the state, and in the progress of society as a thing in itself regardless of the satisfactions of the individuals in it. I believe with all my heart and head that the democratic principles are biologically sound and humanly eugenic, the totalitarian principles unsound and dysgenic. I believe that it is our duty, not as citizens of a democracy but as among the dwindling number of citizens of the world still privileged to live and think as individuals, to oppose the totalitarian fallacy and to maintain the true place of the individual in our social and in our biological philosophy.

The way out of biological-totalitarianism was, for Simpson, found in novel epistemological concepts:

> A newer and, I think, incomparably truer and more profitable point of view is making rapid headway although still far from universal recognition. This is that the group is best definable as a collection of individuals and not as an abstraction of the nonindividual.

This novel political philosophy of biology demanded the end of groups as objects of evolution:

> The group is not an entity in the sense that the individual is an entity. A group achieves adaptation and progresses only in the sense that the

> individuals composing it do so. Satisfaction is an individual compulsion and not a group achievement. Evolution is not a thread on which individuals are strung, but a structure composed of individuals. A species is not a model to which individuals are referred as more or less perfect reproductions, but a defined field of varying individuals.

The "aggregation ethics of the Chicago school" (Mitman, 1992: 164) and the wider notion of group cooperation, fatally associated with political totalitarianism, would not stand long against the hardening synthesis. As Greg Mitman writes:

> David Lack's 1954 book, *The Natural Regulation of Animal Numbers,* signaled an early warning that the population, viewed as a social organism, was dead.... Competition between individuals, rather than cooperation among individuals, became the major force structuring the economy of nature. (1992, 207)

Vero Copner Wynne-Edwards's 1962 book on group selectionism, *Animal Dispersion in Relation to Social Behavior,* was a last gasp, but the shift toward individualism had already consolidated. Against Wynne-Edwards, George C. Williams (1966) persuasively argued that there was no such a thing as the "good of the species."

Following William Hamilton's newly coined concept of "individual fitness" (1964, see Segerstråle, 2013), what mattered to Williams was "the extent to which [an individual organism] contributes genes to later generations of the populations of which it is a member" (1966:97). This "genic view" paved the way for Richard Dawkins's popularization of the "selfish gene," a further paradigm shift (Segerstråle, 2000). Indeed, in the 1960s and 1970s, challenges to the view of the individual as central in evolution (the orthodox modern synthesis view) no longer came from above, the group, but below, the gene (Gould).

However, this 1960s/1970s genic challenge ought to be understood as a radicalization of the individualist one the modern synthesis initiated. The new paradigm implied that apparent group-level phenomena – including pro-social behavior, cooperation, and altruism – could be better explained as a form of genetic investment (Hamilton, 1964) or self-interest (Trivers, 1971). As Herbert Gintis puts it, "The explanatory power of inclusive fitness theory and reciprocal altruism convinced a generation of biologists that what appears to be altruism – personal sacrifice on behalf of others – is really just long-run genetic self-interest" (Gintis, 2006: 106).

But this radicalization of individualism projected now onto the genic level did reflect an ethos unlike that of the modern synthesis. If the synthesis was liberal-democratic with its respect for the agency and autonomy of the individual organism, the new emphasis on selfishness and the transformation of kinship into genic forms of capital investment resonated with the new neoliberal political paradigm that was cementing itself within the global economy in the 1970s, a story that we cannot cover here.

Assessing post-war political biology

It is difficult to deny that something profound and traumatic happened to the politics of biology since the late 1930s and more profoundly after 1945. Eugenics is an exemplar site to look at for these traumatic changes. On one side, there were no longer International Eugenics Conferences, and what remained of the eugenic movement "felt itself constantly under suspicion of sharing the state-interventionist inclinations of the Interwar eugenics movement, which it denied" (Barrett and Kurzman, 2004). Eugenics Journal changed their names, though some of them much later: the *Annals of Eugenics* became the *Annals of Human Genetics* in 1954 and the *Eugenics Quarterly* was rechristened *Journal of Social Biology* in 1969. On the other hand, it is also undeniable that "substantive trajectories" of eugenics did not disappear after 1945; they continued informing the moral economies of post-war debates (Bashford, 2010), not to mention how coerced sterilization continued undisturbed in many Western countries until the 1970s. Especially in the immediacy of 1945 the disappearance of eugenics was in some ways more "cosmetic" than "substantive," in Robert Proctor's terms (1988: 303). After 1945 eugenicists suddenly became population scientists, human geneticists, psychiatrists, sociologists, and anthropologists, but a change of title doesn't imply a change of mind (Kühl, 1994). Typical of these cosmetic changes is Paul Diepgen's *Die Heilkunde und der ärztliche Beruf* that in 1937 celebrated Nazi racial science, while in the 1948 edition suggested that science "serves the entire world and is cosmopolitan" (Proctor, 1988: 304). Major figures of Nazi racial science, from Verschuer to Lenz, were awarded rechristened chairs in human genetics after 1950 (see Müller-Hill 1988, 1998). Immediately after the war the German psychiatrist Karl Bonhoeffer attempted to establish a new eugenics policy (Kühl, 1994: 123). And there was Huxley's explicit reference to eugenics in his 1947 philosophical treatise for UNESCO.

This revision to the conventional account that eugenics ended with the revelations of Nazi horrors comes courtesy of a more recent historiography. The new interpretive school has several strengths. For one, it postpones the moral critique of eugenics to a much later period, after the civil rights movement in America and 1968 in Europe. It also highlights eugenics' persistence in the many sterilization laws that remained on the books until the 1970s, from North America to Europe, and more profoundly in biomedical values that resonated in debates on abortion, population control (Bashford, 2010), and the launch of the Human Genome Project (Duster, 2003).

The new wave of historiography has also challenged another conventional story – the supposed disappearance of scientific racism in the wake of World War II. The teleological shift from race to population could be maintained only "at the cost of historical nuance" (Gannett, 2001:S483). Michelle Brattain argues that the UNESCO project "to dislodge racism [was] equally contingent, opportunistic, political, and grounded in the same social formation as racism itself" (2007). Concerns about the genetics of human variations and human populations thrived long after World War II (Lipphardt, 2012; Lipphardt, 2014; de Chadarevian, 2014). These careful historical studies of race and eugenics confirm the huge gap between, on the one hand, the ideological claims of post-war biology and, on the other, its practices and experiments.

But while I have sympathy for the thesis that eugenics and race persisted after World War II, I am unconvinced by a too-strong continuist view according to which things continued business as usual. The sociopolitical values with which the sciences of human heredity and evolution were aligned, and which they partly produced, were profoundly altered. To pretend that this was merely superficial cover for an unchanged scientific content would be mistaken. Science and society, political and epistemic values, are genuinely coproduced in history. Though prewar concepts remained on the board, the rules of the game had changed. For instance, as Nadia Abu El-Haj recognizes, race has stuck with us since World War II, but it is not "quite the same concept, object, or technology" as it had been (2007).

Arguments about continuity across ages are even more problematic than arguments based on epistemic discontinuities and ruptures. We may even concede, for sake of deconstruction, that Eugen Fischer and Dobzhansky thought in a similar way about race and population (Lipphardt, 2012: S74). But epistemic moments are articulated according to rules deeper than individual speech. Taking note of the four pillars, we see that, starting in the early 1940s, new boundaries came into being. After that, what one could say about race, individual differences, and the politics of biology generally had changed. With a new society, new political epistemological constraints and therefore a new regime of truth came into being. The "continuist" view misses how the modern synthesis and the post-war politics of biology in general shifted the "ground of the debate" (Farber, 2009).

It is interesting to notice the dates of the four conceptual moves I have described in this chapter: Simpson's conference occurred in 1940; Mayr began to promote population thinking in 1942; Dobzhansky and Montagu wrote about escaping from the boundary of nature in 1947; Dobzhansky rethought the importance of genetic pluralism in 1937 and translated this into politics after the war. Dunn claimed that his 1944 radio conferences awoke him to political awareness (in Gayon, 2003). I want to refrain from having an overly deterministic view of culture. However, this clustering of dates speaks to a transition that occurred in a specific political atmosphere. These four pillars can be seen as mileposts indicating not just new science but the beginning of a new political-epistemic discourse. They made possible, and even urgent, certain scientific statements that were (at least some of them) technically possible long before they were uttered but were very much at the margin until the new political milieu demanded them.

These new conceptual pillars have remained robust, and the political shift has held out in the long run. To see how profound and solid were the liberal-democratic foundations of post-war biology, compare two forms of biological determinism, which have much in common epistemologically, but yielded different politics: first, the modernistic ethos of pre-1945 eugenics; second, the conservative discourse of human nature in late-twentieth-century sociobiology and evolutionary psychology, the heretic epigones of neo-Darwinism.

The experiences of sociobiology and especially evolutionary psychology show that even when post-1945 political biology drifted rightward, and even when one of the four pillars (the uniqueness of the human mind or the psychosocial) was undermined, the democratic framework created by the modern synthesis remained stable. Indeed, far from a new eugenics, the rightward turn of political biology has produced a robust theory of human nature opposed to the social engineering ethos typical of eugenics. It was a conservative turn not a eugenic one.

Intermezzo II: conservatism without eugenics: the political philosophy of sociobiology and evolutionary psychology

The first sociobiological writings came as a surprise amid the environmentalist atmosphere of 1970s. Not only had the political terrain moved very much to the left, but sociobiology and later evolutionary psychology were the first schools since 1945 to break with one of the pillars of liberal-democratic biology: the recognition of autonomous human culture and unique human behavior and mental processes.

In a move very close to interwar eugenicists such as psychologist William McDougall (1871–1938) and ethologist Robert Yerkes (1876–1956), and expanding on popular ethological writings of the 1960s, sociobiologists and evolutionary psychologists emphasized the strong connection between animal and human behavior and the dependence of mental mechanisms, including language, on hereditary processes (Gillette, 2007). Against the Dobzhansky-Montagu line, they claimed that a proliferation of instincts, not plasticity, distinguished the human mind. Like 1920s eugenics, 1970s sociobiology and 1990s evolutionary psychology argued that nature prevails over nurture, human culture is genetically determined, and biology should subsume the social sciences.

Even so, and against the predictions of some worried observers, eugenics never reared its ugly head under the guises of sociobiology and, particularly, evolutionary psychology.[1] Given the strength of the democratic framework, even with one of its pillars smashed, and even in a context of the growing reconnection between genetics and human behavior, it was impossible to reopen the eugenic road in any explicit sense, that is at an open ideological level. In fact, the opposite occurred. Sociobiology and evolutionary psychology were undoubtedly conservative responses to the liberal consensus of their time, and in that sense, a challenge to the modern synthesis (or narrow interpretations of some of its assumptions, such as selectionism and adaptationism, see Lewontin, 2007). But their political outcome was anti-eugenic: a theory of universal, biologically hardwired human nature, immovable by social pressure and social engineering.

Although this discourse was aimed primarily at left-wing attempts to manipulate a malleable human nature – the "blank slate" that Steven Pinker criticizes (2002) – it also could have been marshaled against fascism, or as in the case of Fukuyama, genetic engineering (2002). It should be recalled that one of the political roots of the tenacity of human nature was undoubtedly leftist: Chomsky's argument that if humans have no innate structures, then they are easy prey for political manipulation.

> If, in fact, man is an indefinitely malleable, completely plastic being, with no innate structure of mind and no intrinsic needs of a cultural or social character, then he is a fit subject for the "shaping of behavior

> by the state authority, the corporate manager, the technocrat, or the central committee. (*Language and Freedom*, 1970; quoted in Pinker, 2002: 300)

Where the Nazi belief that all traits are heritable meant that a Jew would always be a Jew, no matter how many books he or she wrote on Goethe, in the late twentieth century, the same belief defended the notion that people could not be manipulated for eugenic aims. Sociobiology and evolutionary are perfect examples of how hard-hereditarian scientific programs, based on the primacy of the gene and the preeminence of heritable traits over "acquired" ones, are malleable resources for the political context of the moment – in their case, resources for democracy (in the conservative sense here of a defense from totalitarianism).

True, both sociobiology and evolutionary psychology score poorly on gender equality, and both have circulated Western-centric stereotypes (see Pinker, 2002). But at least racism is kept at bay. Where human nature is universal and fixed since the Pleistocene, alongside a long list of other human universals, few group-based differences are possible (Brown, 1991; Barkow, Cosmides, and Tooby, 1992; Pinker, 2002). "Models of a robust, universal human nature by their very character cannot participate in racist explanations of intergroup differences," Leda Cosmides and John Tooby claim in *The Adapted Mind*, the foundational book of evolutionary psychology (1992: 38).

Relativism, associated with social constructionism and therefore environmentalism, was now the divisive enemy leading to the cleavage of humankind rather than its unity – another curious reversal of the political function of hard heredity. Evolutionary psychology's view that cultural variations (or "environmental cues") are merely superficial, while the inner, evolved core of the human mind depends on the "psychic unity of humankind" (Barkow, Cosmides, and Tooby, 1992), can even be seen as a radicalization of the first UNESCO statement: a common *biological* human nature as a basis for universal brotherhood. When Cosmides and Tooby claim, "By virtue of being members of the human species, all humans are expected to have the same adaptive mechanisms," they are in a way channeling the UNESCO spirit.

This is not to deny the potential for racism in both sociobiology and evolutionary psychology, and xenophobic groups have exploited the rhetoric of both movements. In particular sociobiology's naturalization of the ethnocentric instinct (Wilson, 1978: 119) was an easy inspiration for racists. But Wilson and Pinker explicitly omitted race from their agendas (see Jumonville, 2002; Pinker, 2002: 144). In a 2001 interview Pinker declared:

> If our society did not divide people by race then the question of racial differences would be too scientifically boring for anyone to bother with. Races are biologically superficial, and they tie in to no real theory of how we evolved, so there is no coherent explanation as to why races should differ biologically. I don't think scientists would even be interested in the question. (Shermer, 2001)

Epistemological critiques of evolutionary psychology, such as David Buller's *Adapting Minds: Evolutionary Psychology and the Persistent Quest for Human Nature* (2005), have emphasized polymorphism and differences in human psychological attitudes. Of course, the evolutionary basis of evolutionary psychology has always been thin (ibid.). Moreover, it implies deep assumptions of what counts as normal human nature, which may have profoundly exclusionary implications. But certainly the debate had shifted a lot from the period when supporting differences and polymorphism and even the idea of "a variety of adaptational and genetic 'natures' in human populations" (Buller, 2006: 424) was seen as hatching potential for racism. Now it has in fact become a progressive critique of conservative evolutionary psychology (perhaps because the universalist framework has become stabilized enough at a different level).

Buller may put "nature" in scare quotes, but for Pinker, there is no question that human nature is real, fixed, and of tremendous political importance. Building on the work of conservative political theorist Thomas Sowell, Pinker counters a utopian vision whereby "human nature changes with social circumstances" with his tragic, intractable one, in which human nature cannot be altered by social pressure (2002: 288–289). The utopianism that Pinker's innatism attacks looks much like the bio-utopianism of eugenics. The political implications are clear. "Human nature is the reason we do not surrender our freedom to behavioral engineers."

> Inborn human desires are a nuisance to those with utopian and totalitarian visions, which often amount to the same thing. What stands in the way of most utopias is not pestilence and drought but human behavior.... The Marxist utopians of the twentieth-century, as we saw, needed a tabula rasa free of selfishness and family ties and used totalitarian measures to scrape the tablets clean or start over with new ones. (2002: 169)

Pinker cleverly reverses the "moral appeal" of environmentalism (the doctrine of the blank slate) as the basis of political emancipation:

> The Blank Slate was an attractive vision. It promised to make racism, sexism, and class prejudice factually untenable. It appeared to be a bulwark against the kind of thinking that led to ethnic genocide.... But the Blank Slate, had, and has, a dark side. The vacuum that it posited in human nature was eagerly filled by totalitarian regimes, and it did nothing to prevent their genocides.

This is a typical neoconservative move, repeated throughout *The Blank Slate*:

> If people are assumed to start out identical but some end up wealthier than others, observers may conclude that the wealthier ones must be more rapacious. And as the diagnosis slides from talent to sin, the remedy can shift from retribution to vengeance. Many atrocities of the twentieth century were committed in the name of egalitarianism. (2002: 152)

My concern is not with the historical plausibility of this view. What is significant is how a scientific belief in nativism, inborn factors, and innateness, as proxy for hard-hereditarianism especially of the right-wing kind, was turned in the late twentieth century into a political tool of the promotion and defense of a particular conception of Western democracy.

Fukuyama's usage of human nature as described by an "increasing body of evidence coming out of the life sciences" is similar. Like Pinker, Fukuyama believes, "The standard social-science model is inadequate" and "human beings are born with pre-existing cognitive structures and age-specific capabilities for learning that lead them naturally into society (1999: 155).

For Fukuyama, human nature supports the hope of countering the "great disruption" occurring in American society in the 1960s and 1970s as an effect of a shift in moral values as well as various wrongly conceived social engineering projects. Social order and social capital can be reconstructed, Fukuyama claims, thanks to the "genetic basis" of cooperative and pro-social behaviors (1999: 168). The "tenacity" of a genetically based human nature is again enlisted as defense of democracy, and a tool to remake society. As Fukuyama writes in *Our Posthuman Future* (2002):

> For while human behavior is plastic and variable, it is not infinitely so; at a certain point deeply rooted natural instincts and patterns of behavior reassert themselves to undermine the social engineer's best-laid plans.

Fukuyama uses human biology to set "the framework for whole human politics" (Fukuyama, 2011). So do the Darwinian conservative Larry Arnhart (1998, 2005) and Melvin Konner, via Darwinian constraints of human behaviors (2002). In fields ranging from neuroscience to primatology, Michael Gazzaniga, Marc Hauser, Frans de Waal, Jonathan Haidt, and others propose a latter-day moralized version of human nature (Meloni, 2012, 2013). In all these discourses, a well-delimited biological nature, human or not, is the source of a profound political hope. All of these authors promote concepts of heredity with which interwar eugenicists would be comfortable, but none of them proposes eugenics itself.

Comparing the modernistic ethos of eugenics with the conservatism of human nature in post-1970s sociobiology and evolutionary psychology confirms that similar scientific assumptions can be turned toward different ideological goals: engineering of biologically based human nature for the improvement of the race versus biologically universal human nature in opposition to engineering. Eugenics and sociobiology/evolutionary psychology both assert the ontological primacy of heredity and both flatten the human to its biological dimension, but the political outcomes of these two versions of political biology are incommensurable: utopian (or dystopian) and based on social engineering in the first case, tragic or conservative in the second.

Scientific views produce different political agendas depending on the normative systems in which they circulate. Post-war democratic thought is such a system, and it has tenaciously prevented certain scientific statements achieving policy voice, and completely silenced others. The stability of democratic biology can be deduced from the fact that the radical hereditarianism of sociobiology and evolutionary psychology, so scientifically similar to interwar eugenics, has been articulated in defense of Western democracy, albeit from a conservative, or libertarian, perspective. However, as this book will argue next, now for the first time some of the key assumptions of post-war democratic biology are faltering under the weight of a new postgenomic view of biological processes, heredity in particular.

7
Welcome to Postgenomics: Reactive Genomes, Epigenetics, and the Rebirth of Soft Heredity

Welcome to postgenomics[1]

From the modern to the extended synthesis (or a new one?)

Over the last two decades, profound conceptual novelties have influenced our understanding of evolutionary thinking. The implications for the notion of heredity have been profound. In 2008, when globally influential philosophers of biology met in Austria to reflect on the status of evolutionary theory 150 years after Darwin's *Origin of Species*, it was clear that the modern synthesis of Darwinism and genetics needed revision (Pennisi, 2008). The outcome of that meeting was Massimo Pigliucci and Gerd B. Müller's *Evolution: the Extended Synthesis* (2010; see also Pigliucci, 2009), which notes

> Evolutionary theory, as practiced today, includes a considerable number of concepts that were not part of the foundational structure of the Modern Synthesis... for several years now dissenters from diverse fields of biology have been questioning aspects of the Modern Synthesis, and pivotal novel concepts have been elaborated that extend beyond its original scope.

According to Pigliucci and Müller, these dissenting voices are calling "for an expansion of the Modern Synthesis" (3). They argue that "individual tenets of the Modern Synthesis can be modified, or even rejected, without generating a fundamental crisis in the structure of evolutionary theory – just as the Modern Synthesis itself improved upon but did not cause the rejection of either Darwinism or neo-Darwinism" (8).

Others dispense with any such accommodation. Michael Rose and Todd Oakley suggest that the "new biology" has so "severely challenged the assumptions of the 'Modern Synthesis'" (2007) that evolutionary thinking must be radically restructured (see also Gilbert and Epel, 2009; Jablonka and Lamb, 2014; Danchin, et al., 2011; Danchin, 2013; Koonin, 2012; Mesoudi et al., 2013). David Depew and Bruce Weber, two other distinguished philosophers of biology, compare the modern synthesis to Newtonian physics: valid in its own domain but no longer able to claim general validity (2011).

Leaving aside this debate about reform or subversion of the synthesis (see also Depew, 2013), what is important to recognize is that over the past few decades, diverse movements have challenged some of the pillars of twentieth-century biology. These movements include developmental systems theory (DST); niche-construction; evolutionary developmental biology, or evo-devo, which incorporates generative and developmental mechanisms; and ecological developmental biology, or eco-devo.

The research claims, for example, that biological information is not localized in the genome alone but is dispersed thorough the whole organism (Oyama, Griffiths, and Gray, 2001; Robert, 2004) and that its flux is bidirectional, not just from genes to proteins per Crick's dogma of molecular biology (Griffiths and Stotz, 2013). New theories suggest that a single genotype can produce different phenotypes in response to different environmental triggers and therefore that a phenotype cannot be "mechanistically deduced, even if we possess a complete DNA sequence of its genome" (Schlichting and Pigliucci, 1998; Pigliucci, 2001; West-Eberhard, 2003; Robert, 2004). We have learned that the organism does not adjust passively to the environment but rather contributes to the "construction of its niche which, in turn, can re-orientate the evolutionary process" (Odling-Smee, Laland, and Feldman, 2003; see also Lewontin, 1983). The phenotype is also extricated from the passive position Weismann assigned it (Oyama, 2000a and b; Griesemer, 2002; Meloni and Testa, 2014). The organism can "modify significant sources of natural selection ... thereby codirecting subsequent biological evolution" (Laland, Odling-Smee, Laland, and Feldman, 2000). And development does not appear separated from heredity and therefore cannot be bracketed away when understanding evolution (see Amundson, 2005; Sultan, 2007; Müller, 2010).

All these ideas imply an enlarged evolutionary role for the environment – that "environmental factors can elicit innovation not via natural selection but through their direct influence on developmental systems" (Müller, 2010: West-Eberhard, 2003). Thus the new biology also confronts the "gene centrism" of the modern synthesis, its exclusive focus "on the gene as the sole agent of variation and unit of inheritance" (Pigliucci and Müller, 2010, 14). Ironically, the crisis of gene centrism is the result of accelerated research since the conclusion of the Human Genome Project (HGP). As Evelyn Fox Keller noticed fifteen years ago, advances in genomics show "the ever widening gaps between our starting assumptions and the actual data that the new molecular tools are now making available" (2000: 8).

Even before the completion of the HGP, it had become clear that developments in molecular biology – the discipline that was supposed to offer a realist and reductionist view of the material units of life – have made it "impossible to think of the gene as a continuous piece of DNA matter collinear with a piece of protein matter" (Rheinberger, 2003: 232). Since the completion of the HGP, the gene has come under yet more scrutiny, its role as a particulate and autonomous agent determining traits and developmental processes becoming more difficult to reconcile with scientific evidence. What Brendan Maher calls "missing heritability" – the lack of correlation between genetic variants and common traits or complex diseases – illustrates the unexpected complexity of developmental space between genotype and phenotype (Maher, 2008; Eichler et al., 2010).

This growing complexity has cast further doubt on what a gene is and does (Moss, 2003). The image of the gene performing just one job – coding for proteins – has been radically overturned (Portin, 2009; Keller, 2012, 2014; Griffiths ad Stotz, 2013). Not only does a small percentage of the genome (less than 2%) act according to the classical definition of the gene as a protein-coding sequence, but most of the non-protein coding DNA is far from useless. What was supposed to be "junk DNA" turns out to play an important regulatory function it could not have had in the "gene-centric view" (Encode, 2007, 2012; Pennisi, 2012). On the gene-centric view, genomes are "littered with nonfunctional pseudogenes, faulty duplicates of functional genes that do nothing, while their functional cousins... get on with their business in a different part of the same genome" (Dawkins, 2010). But we are now realizing that the "junk" is actually "a major player in many of the processes that shape the genome and control the activity of its genes" (Biémont and Vieira, 2006; Mattick, 2003, 2004; Pink et al., 2011). As Petter Portin writes:

> The structural boundaries of the gene as the unit of transcription are far from clear... the human genome is pervasively transcribed from both DNA strands, such that the majority of its bases can be found

> in primary transcripts, including non-protein-coding transcripts, and those that extensively overlap one another. The complexity of the transcription of protein-coding and non-coding RNA sequences is evident: transcripts may be derived from either of both DNA strands, and they may be overlapping and interlaced, and the transcripts can even use the same coding sequences. (2009: 113)

In this context, it is becoming harder to maintain a particulate notion of the gene. It is more accurate to think of genes in a deflationary way as "complex spatially discontinuous objects – composite rather than unitary objects" (Barnes and Dupré, 2008, 59) embedded in a regulatory network with distributed agency and specificity (Griffiths and Stotz, 2013). An important part of this regulatory network is involved in responding to environmental signals, which can originate in the cellular environment around the DNA, the entire organism, and, in the case of human beings, their social and cultural dynamics.

These exciting new concepts comprise the terrain of the postgenomic age.

Postgenomics as a different style of reasoning

Postgenomics is a term in flux, trendy but difficult to define. The first uses of the term came in the late 1990s during a series of preparatory conferences for the launch of the Human Genome Project (Richardson and Stevens, 2015).

Since then, the term has appeared in the titles of several publications, but has never been clearly explained or differentiated from genomics. This usage may imply that the meaning of postgenomics is so obvious that there is no reason to say more about it. But given the growing conceptual turbulence in evolutionary thinking, much is to be gained by understanding postgenomics deeply, as a "style of reasoning" (Hacking, 2002) different from its predecessors.

There are at least five parallel ways of understanding postgenomics. These are not alternatives but rather form a network of coexisting meanings, epistemologies, and practices (El Hai, 2007; Richardson and Stevens, 2015).

The first meaning of "postgenomics" is temporal. Here *post* just means *after*, so that postgenomics indicates simply "the period that followed the publication of the draft human genome sequence in 2001" (Griffiths and Stotz, 2013). Postgenomics "signposts the most recent period in the history of the life-sciences" but does not imply fundamental rupture (Ankeny and Leonelli, 2015).

A second meaning points to the emergence of a "unifying framework for biological knowledge" that brings together the whole new set of approaches, dubbed the "-omics" that extend the existing genomic programs and paradigms across the many subfields of the life sciences" (Richardson and Stevens, 2015; Sunder Rajan and Leonelli, 2013). Today, the -omics include not only the genome but also the microbiome, transcriptome, nutrigenome, exposome, proteome, metagenome, and as we shall see later the epigenome. -Omics, a Greek suffix indicating "wholeness" or "collectivity of the units in the stem" (Lederberg and Cray, 2001; Yadav, 2007), here refers not only to a trend toward studying phenomena (genes, proteins) in their wholeness and interaction but also at a bigger infrastructural scale of program and financial investment.

The third meaning concerns specifically the infrastructural and technological dimension of the studies of genes, such as new sequencing technologies whereby large-scale genomic, metabolomic, or exposomic maps and databases are obtained (Rheinberger, 2013; Mackenzie, 2015; Ankeny and Leonelli, 2015). In this novel data-driven setting (Strasser, 2012), the environment is understood as a signal, (Landecker, 2016) which can be encompassed in genome-friendly, code-compatible digital representations (Meloni and Testa, 2014).

A fourth meaning of postgenomics has to do with a new political and moral economy, largely coinciding with the neoliberal era. Social scientists have argued that "postgenomics is a distinctly neoliberal science in terms of its economic – its commercial structure" as well as "part and parcel of the domain of speculative finance" (Abu El-Haj, 2007; see also Thacker, 2005). The moral economy of postgenomics is also neoliberal: it focuses on individual risk (Sunder Rajan, 2005; Abu El-Haj, 2007; Rose, 2007) and self-optimization (Prainsack, 2015; Mansfield, 2012; see also Novas & Rose, 2005). Postgenomics is part of a new cycle of hype and promises (Pickersgill et al., 2013) and equipped with a particular affective economy (or assemblages, see Fortun, 2015).

The fifth understanding of postgenomics is not just chronological, infrastructural, technical, or politico-economic, but conceptual. Postgenomics can be understood as the emergence of "unanticipated levels of biological complexity" in our understanding of the genome (Keller, 2015). This has led to radically rethinking the ontology of the genome as "pre-existing developmental processes" (Robert, 2004: 74) and even dismissing its role as the prime mover in biological processes

(Stotz et al., 2006; Barnes and Dupré, 2008; Griffiths and Stotz, 2013). Postgenomics in this sense can be taken as an invitation to go *beyond the genome* as we have known it. As Karola Stotz writes (2006):

> For the largest part of the past century we came to see genes as a material unit with structural stability and identity, with functional specificity by means of their template capacities that encode information, and with intergenerational memory; we came to see genes as the designator of life and the site of agency and even mentality (in containing a plan or program for and asserting control over developmental processes). In the postgenomic era, however, there is no DNA sequence that exhibits any or all of these traits without the help of an extensive and complex developmental machinery.

The "postgenomic genome" (Keller, 2015) exists within this broader developmental architecture.

De-centering and temporalizing the genome

The postgenomic genome is a flexible and blurred entity whose spatial and temporal dimension have been extended and emphasized to an unprecedented degree (Keller, 2015), an entity that "is far more fluid and responsive to the environment than previously supposed" (Jablonka and Lamb, 1995). An excellent way to capture these novelties is to think of the genome as a "vast reactive system" (Keller, 2012), a mechanism "for regulating the production of specific proteins in response to the constantly changing signals it receives from its environment" (Keller, 2014, 2427). Following Nobel Laureate Barbara McClintock, we might conceive of the genome as a "highly sensitive organ" (1984; see Keller, 1983). When first formulated in her Nobel Prize speech, the idea seemed quite eccentric, but it has been fully vindicated in recent times (Jablonka and Lamb, 2014).

Spatially, we now understand that the genome's borders with the environment are porous – indeed, almost impossible to establish. "The idea of a distinct molecular gene with clearly defined boundaries" has become untenable (Griffiths and Stotz, 2013: 68). Far from being a separate and autonomous agent, the postgenomic genome is embedded in a broader regulatory architecture that "extends outside the organism into the developmental niche." This is why "rather than looking for causes in DNA sequence information," the focus in postgenomic research "has shifted towards how sequences are used in a transient and flexible way by the varied mechanisms which control gene expression" (ibid.).

Temporally also, the postgenomic genome is radically new (Lappé and Landecker, 2015). This genome is subject to time and biography and even, according to Martine Lappé and Hannah Landecker, possesses "an early life and an old age, and to a more limited degree, an adolescence, middle age, and other stages." Lifetime experiences "impinge on the formerly implacable and sequestered genome." As they claim, "developments in chromatin biology have provided the basis for this genomic embodiment of experience and exposure." This is an impressive shift from the picture of the genome "that came into being through the massive sequencing efforts of the 1990s and 2000s in which genomes were understood as the same in every cell of the body for all of that body's life" (ibid.).

This novel spatial and temporal condition helps to illustrate why postgenomic views have to be *contrasted* with the genetic and genomic paradigm. Though postgenomics is often a matter of business-as-usual "gene hunting" (Bliss, 2015), though it has been criticized for reinforcing "discourses about the power of genes" (Richardson and Stevens, 2015; see also Panofsky, 2015 for behavioral genetics; and Waggoner and Uller, 2015), and though some scientists insist epigenetics offers "little more than another layer of DNA related activity that needs to be taken into account when analysing gene expression," (Niewöhner, 2011) my claim is that we are seeing a real rupture. Postgenomics should not be conceived as a mere prolongation of what came before (Charney, 2012).

There is no better illustration of the discontinuity of postgenomics than the destabilizing effect it has had on the central dogma of molecular biology. We have already seen how the basic notion of the dogma – information going one way from DNA to protein and never in reverse – (1958: 153; see also Crick, 1970; Olby, 1970; Strasser, 2006 for the broader meaning assumed by the dogma, and Morange, 2008 for an assessment after 50 years) was constructed to disprove any direct or formative influence of the environment on the gene. No alteration in protein sequence could cause subsequent alterations in nucleic acid sequence (Olby, 1970). The line of causality was clear: "DNA makes RNA, RNA makes proteins, and proteins make us" (Keller, 1996: 18; see also Watson quoted in Strasser, 2006).

In retrospect one could say that the dogma has always been unstable. Since peak consolidation in the 1960s (Strasser, 2006), it has undergone several theoretical and empirical challenges. Keller (1992) has shown how attempts to jeopardize the dogma or bring attention to ambiguous findings were neutralized by the scientific community, which played on the semantic instability of scientific terminology (for instance: "directed" and "spontaneous" mutations) and reframed the boundaries of the dogma to make it compatible with new claims. However, what were once minor wrinkles have become "major chasms" (Keller, 2000: 55 and ff.), and it is generally believed that the dogma as an absolute principle is today "invalid" (Koonin, 2012). While it is true that *strictu senso* there is no reverse translation from protein sequence to DNA (in this sense the dogma does hold), the broader view of the dogma according to which there is no transfer of information from phenotype to genotype and also no transfer from phenotype to phenotype (e.g. protein to protein) does no longer hold. Plenty of molecular processes – such as direct DNA translation to protein, transfer of information from protein sequences to the genome, genetic assimilation of prion-dependent phenotypic heredity, reverse transcription, non-protein-coding RNA (ncRNA) transcription, and transplicing – have disproved molecular biology's "linear logic" (respectively in Pigliucci and Muller, 2010; Koonin, 2012; Brosius, 2003; Mattick, 2003; see also Dupré, 2010; Hayden, 2011; Charney, 2012; Jablonka and Lamb, 2014). Today we know that biological information goes in both directions and is not only contained in DNA sequences: "instead of a linear flow of information from the DNA sequence to its product, information is created by and distributed throughout the whole developmental system" (Stotz, 2006; Griffiths and Stotz, 2013).

We must dispense with germplasm sequestration as imagined by Weismann and made molecular by Crick. This is a massive change. After all, the dogma's subtext was that there was "no direct route by which the environment could imprint on DNA" (Lappé and Landecker, 2015). The hardness of the hereditary material was secured (Olby, 1970). As Eva Jablonka and Marion Lamb write, "Because of the central dogma there was no way in which induced phenotypic changes could have any effects on the genetic material" (2014). Epigenetics, a milestone in the postgenomic landscape, shows how far we have already moved from molecular biology's rigid assumptions. With epigenetics we know that we do not need reverse translation to transmit information from phenotype to genotype: "What changes is which genes are switched on and which are switched off. It is the *amounts* of the various proteins, not their sequences, that are altered. Backtranslation is irrelevant to transmitting such alterations" (Jablonka and Lamb, 2014).

Enter epigenetics

Epigenetics is a perfect incarnation of postgenomics and a high-resolution theoretical spyglass through which to see the changing thought-style, and possibly ethos, of the biosciences in the early twenty-first century. More than fifty years after epigenetics was introduced, it has become a buzzword (Jablonka and Raz, 2009: 131); the growth of publications in the field in the last decade certifies the epidemic of epigenetics (Haig, 2012; Jirtle, 2012).

I do not intend to oversell the conceptual and evidentiary strength of a discipline still as embryonic, multiple, and contested as molecular epigenetics. Many questions in epigenetics remain highly controversial, and we should continue to be cautious about its relevance, especially for humans (Feil and Fraga, 2012; Heard and Martiennsen, 2014). Moreover, the notion of epigenetics is elusive and plastic, (Morange, 2002; Bird, 2007; Ptashne, 2007; Dupré, 2012; Griffiths and Stotz, 2013). Despite – or, more likely, because of – this semantic ambiguity, epigenetics prospers as a scientific and social phenomenon in need of careful scrutiny (Meloni and Testa, 2014). The genealogy of epigenetics in biological thought is complex, and its current molecular crystallization is the result of a series of important conceptual shifts (Jablonka and Lamb, 2002; Haig, 2012; Griffiths and Stotz, 2013). This is not a straightforward matter, and I will therefore approach it with care.

"Epigenetics" was coined by embryologist and developmental biologist C. H. Waddington (1905–1975) in the 1940s as a neologism "derived from the Aristotelian word 'epigenesis' which had more or less passed into disuse" (1957: not then, as in the popular but mistaken version in which epi means "above" the gene) to specify, in a broad non-molecular sense, the "whole complex of developmental processes" that connects genotype and phenotype (1942 reprinted in 2012). According to Waddington's 1942 definition, epigenetics explored the unfolding of genetic material into a final phenotype. A later definition saw it as "the branch of biology which studies the causal interactions between genes and their products which bring the phenotype into being" (Waddington, 1968; see Holliday, 1990, 2006; see Jablonka and Lamb, 2002).

The present understanding of epigenetics, however, is more influenced by a second, narrower tradition (Griffiths and Stotz, 2013). This tradition originates with David Nanney's 1958 paper "Epigenetic Control Systems" and refers to a second, non-genetic system, operating at the cellular level, which regulates gene expression (Nanney, 1958; see Haig, 2012, Griffiths and Stotz, 2013). Thus it is probably more correct to call contemporary epigenetics "molecular epigenetics," to differentiate it from Waddington's broader sense and from the developmentalist-embryological tradition in which the term was first conceived. That said, the two meanings are not in principle irreconcilable, as they both emphasize the context – either molecular or at the level of the organism – where genetic functioning takes place (Hallgrímsson and Hall, 2011).

In the currently mainstream molecular sense, a standard definition of epigenetics is "the study of changes in gene function that are mitotically and/or meiotically heritable and that do not entail a change in the sequence of DNA" (Armstrong, 2014). In less technical books, epigenetics is called the study of all the "long-term alterations of DNA that don't involve changes in the DNA sequence itself" (Francis, 2011: X). In a broader but still negative form, epigenetics can be defined as any "phenotypic variation that is not attributable to genetic variation" (Haig, 2012: 15). A rare positive definition calls molecular epigenetics "the active perpetuation of local chromatin states" (Bird & Macleod, 2004, quoted in Richards, 2006: 395) or the self-perpetuation of gene expressions "in the absence of the original signal that caused them" (Dulac, 2010: 729). Amid proliferating–omics, epigenetics can be also seen as the study of the epigenome, the set of potentially "heritable changes in gene expressions that occur in the absence of changes to the DNA sequence itself" (Dolinoy and Jirtle, 2008).

The preferred recourse to a negative definition – heritability *without* altering the DNA – not only reflects the uncertainty surrounding the range and stability of epigenetic mutations, but also and more importantly the difficulties of conceptualizing epigenetics beyond a gene-centric view of heredity and phenotypic development.

Mechanisms

To understand epigenetics it is a good idea to start from the biochemical basis of the process.

In eukaryotic cells, DNA is tightly wrapped into chromatin, and modifications of the chromatin structure can affect DNA expression (Meaney and Szyf, 2005; Feil and Fraga, 2012). DNA methylation – a surprisingly simple chromatin modification, which involves merely adding one carbon and three hydrogen atoms to a receptive site of the DNA – is the most recognized mechanism of epigenetic mutations. Studied since the late 1960s (see Holliday, 2006, see however Daxinger and Whitelaw, 2012) it results in chromatin de-activation and inhibition of gene transcription. Methylation works therefore as a sort of "physical barrier to transcription factors" (Gluckman et al., 2011) and is regulated by nutritional and environmental factors, especially during early phases of development (Dolinoy and Jirtle, 2008). Further epigenetic mechanisms include histone modification, histone protein modifications, and regulation by non-coding RNA, although these are considered less stable than methylation patterns.

In evolutionary terms, epigenetic changes are not biological anomalies but fundamental to developmental plasticity, the "intermediate process" by which a "fixed genome" can respond in a dynamic way to a changing environment and produce different phenotypes from a single genome (Meaney and Szyf, 2005; cfr. also Robert, 2004; Gluckman et al., 2009, 2011). Unsurprisingly, then, there are several examples of visible phenotypic changes in the animal kingdom driven by changes in methylation patterns as a result of varying environmental exposure and nutritional input. For instance, genetically identical honeybee (*apis mellifera*) larvae following different feeding regimens (royal jelly versus less rich food) produce different adult phenotypes, from sterile worker to fertile queen (Kucharski et al., 2008). Nutrition also affects the agouti gene in mice, producing in phenotypic changes in offspring. Exposing a pregnant agouti mouse to a low-methyl diet results in hypomethylation and enhanced expression of the promoter of the agouti gene. Offspring are no longer slim and brown, but yellow, fat, and prone to diabetes (Waterland and Jirtle, 2003). And moving from nutritional to behavioral exposures, Michael Meaney's group study shows that grooming behavior in rats alters the methylation patterns of the promoter of the glucocorticoid receptor in pups (Weaver et al., 2004; see an updated review in Lutz and Turecki, 2014).

Consequences for heredity

One of the most hotly debated issues surrounding epigenetic mechanisms is their transgenerational stability (Richards, 2006; Jablonka and Raz, 2009; Daxinger and Whitelaw, 2010, 2012; Lim and Brunet, 2013; Grossniklaus et al., 2013; Heard and Martienssen, 2014). Received wisdom is that, in mammalian development, epigenetic marks are reset at each generation and therefore incapable of sustaining transgenerational phenotypic changes. But new studies are challenging this received view of inheritance (Anway et al., 2005; Rassoulzadegan et al., 2006; Hitchins et al., 2007; Wagner et al., 2008; Franklin et al., 2010; Saavedra-Rodrıguez and Feig, 2013) and pointing at environmental effects lasting on up to four future generations via epigenetic mechanisms (Whereas intergenerational variations are passed from parents to offspring, transgenerational ones continue for multiple generations.). This can happen in either of two ways. First, by germline inheritance, where the epigenetic signature is not entirely cleared in gametogenesis and can be transmitted through the germline (Chong and Whitelaw, 2004; see Anway et al., 2005). Second, by non-germline, experience-dependent epigenetic inheritance, where the epigenetic signature is reestablished in each successive generation by the reoccurrence of the "behaviour or environment that induces the mark." This is also known as "niche recreation" (Gluckman et al., 2011; Champagne, 2008, 2013; Champagne and Curley, 2008; Danchin et al., 2011).

Epidemiological studies now use epigenetic mechanisms in humans to explain the intergenerational and possibly transgenerational effects of chronic disease[2]. Epidemiological and medical studies have often looked at the "intergenerational transmission of programmed effects" (Drake and Liu, 2010). In the mid-1980s epidemiologist and pediatrician Irving Emanuel proposed the *intergenerational influences hypothesis* (IIH) to account for the way in which "exposures, and environments experienced by one generation [related] to the health, growth, and development of the next generation." Much of the evidence of intergenerational factors, however, was at that point of "a largely indirect nature" (Emanuel, 1986: 35). It has been only in the last two decades that increasing numbers of studies have expanded and tested what was until then the "speculative" intergenerational hypothesis, bringing an entirely new dimension to it.[3] The most famous concerns the effects of prenatal exposure to famine during the Dutch *Hongerwinter* (Hunger Winter) of 1944–1945, when the Germans occupied West Holland (Heijmans et al., 2008; Painter et al., 2008). While the Nazis blocked food transport, caloric intake plummeted below 800 calories per day. The shortage lasted nearly half a year and left profound biological scars: babies born to mothers who experienced famine during the last trimester of pregnancy were significantly smaller than babies born after the country was liberated and displayed several metabolic and behavioral problems, from obesity to mental illness. Effects of the Dutch famine have been recorded up to sixty years later and extend beyond the offspring of those exposed, apparently to the second generation, so far. This lingering impact is correlated with significant changes in methylation (Painter et al., 2008).

Another important line of research investigates links between descendants' health and longevity and ancestors' nutrition and lifestyle. Drawing from records kept by the isolated Northern Swedish town of Överkalix, where three cohorts (born in 1890, 1905, and 1920) were followed until death in 1995, these studies show that ancestors' access to food correlates with descendants' longevity (Bygren et al., 2001) and susceptibility to cardiovascular disease (Kaaty et al., 2002). Researchers have also found transgenerational effects (on the male line) of smoking before puberty when the sperm is in formation (Pembrey et al., 2006). In all of the Överkalix cohort studies, epigenetic changes appear to be the key mechanism for this perpetuation of environmental effects. Interestingly, here it is not only the matrilinear side to be emphasized as vehicle of epigenetic mutations as in the Dutch cohort.

Probably no other topic is so controversial today as epigenetic inter- and transgenerational inheritance in humans. The scientifically unstable status of the topic has been clearly emphasized not only by sceptical scientists but also by social scientists (Tolwinski, 2013; Pickersgill, 2016). There is still a "long way to go" to fully understand "the involvement of epigenetics in environmentally triggered phenotypes and diseases" (Feil and Fraga, 2012; see also Heard and Martienssen, 2014), but this research suggests significant consequences for the notion of biological heredity. What we already know offers reason enough to rethink the investment in hard heredity and the division of the human world into biological and social domains characteristic of the hard-heredity framework.

By challenging the idea that heredity is the mere transmission of nuclear DNA, epigenetics has opened the door to an extended view of inheritance in which information is transferred from one generation to the next by many interacting inheritance systems (Jablonka and Lamb, 2014). Epigenetic variations act as a parallel inheritance system through which the organism can respond more rapidly and flexibly to environmental cues by transmitting to cell lineages different "interpretations" of DNA information (ibid.). Thus it is no longer just the DNA sequence that is transferred but the whole "developmental niche" (Stotz, 2008, 2010), "the set of environmental and social legacies that make possible the regulated expression of the genome during the life cycle of the organism" (Griffiths and Stotz, 2013: 110). This means that environmental and social factors, not only genes, "carry information in development" (ibid.: 179).

Epigenetics is also providing "candidate mechanisms" (Kappeler and Meaney, 2010; see also Danchin et al., 2011) for parental effects, a key phenomenon of developmental plasticity whereby exposures in one generation to certain environmental states (nutrition, toxins, etc.) can affect the next generation's phenotypes without affecting their genotypes (Badyaev & Uller, 2009; Danchin et al., 2011). Maternal effects, in particular, are critical, enabling "the mother to adjust the phenotype of offspring in response to the environment she inhabits and, in doing so, in effect transmit to them information about the environment they will inhabit" (Charney, 2012: 353).

By inviting us to consider the possibility that "heredity involves more than genes" and that "new inherited variations...arise as a direct, and sometimes directed, response to environmental challenge," epigenetics comes close to Lamarckian ideas of soft inheritance (Jablonka and Lamb, 1995, 2014; Champagne, 2010; Gissis and Jablonka, 2011; Richards, 2006; Jablonka and Lamb, 2008; Jablonka and Raz, 2009; Bonduriansky, 2012). But is it misleading to speak of renewed soft heredity? For one thing, the notion was coined by hard-hereditarian authors (Darlington, 1959, Mayr, 1980: see Gissis and Jablonka, 2011) to gather under one umbrella all that was considered wrong or outdated in views of heredity. Second, and more important, talk of soft heredity may give the impression of a return to pre-genetic views of heredity. However, the point is not to give up on genetics but to gain insights from it. And these insights show how the hereditary material is regulated by "contextual factors" that "influence the development of all our characteristics" (Moore, 2015). It may be more accurate to speak not of soft heredity – with its historical baggage – but of inclusive inheritance (Danchin et al., 2011), exogenetic inheritance, (Griffiths and Stotz, 2013) or, in more popular terms, heredity 2.0 (Meloni, 2015a and b).

To recap, there are, allegedly, seven crucial differences between epigenetic mechanisms and genetic ones:

Table 7.1 Differences between epigenetic mutations and genetic mutations

Epigenetic mutations	**Genetic mutations**
1) Sensitive to the environment in the short term – A mechanism for flexible and dynamic responses to signals from a changing environment – Directed variation as a consequence of a "specific environmental agent inducing specific and predictable heritable changes" (Jablonka and Lamb, 1995) – Variations are nonrandom but not necessarily adaptive (Jablonka and Lamb, 1995)	1) Unresponsive to the environment in the short term – Mutations are accidental, e.g., due to exposure to x-rays or nuclear radiation – Mutations are random: "independent of selective pressure" (Rando & Verstrepen, 2007), only due "to the imperfections of the copy-system or to non-directed effects of environmental factors" (Jablonka and Lamb, 1995)
2) Potentially reversibile during the lifetime through practice or therapy, e.g., pharmacological intervention (Szyf, 2001, 2009a and b)	2) Affected by lifestyle only over a long evolutionary timescale, as in gene-culture coevolution, e.g., lactase persistence in human populations that consume large amounts of dairy products.

Continued

Table 7.1 Continued

3) Sensitive to *in utero* and perinatal events – Early life experiences and environment embeded in the genome via epigenetic mechanisms (Champagne and Curley, 2009; Turecki et al., 2012; Szyf and Bick, 2013). – Epigenetic responses to early life adversity have genome-wide effects (McGowan et al., 2011)	3) Insensitive to in utero and post-natal events – No link between early life events and genetic sequences, except as caused by early exposure to random mutations
4) Short-term adaptive flexibility – Responsive to transient ecological challenges – Stable enough to allow transmission over a limited number of generations	4) Long-term adaptive flexibility – Thousands of years pass before changes are fixed in the genetic pool
5) Tissue and cell specificity – Different cells have different epigenetic marks – One body, many epigenomes (Wade, 2009)	5) One body one genome
6) Time dependency – Epigenetic marks vary depending on the when a sample is taken – Age-related changes are evident in epigenetic marks	6) Time independence – DNA is unchanging throughout the lifespan
7) Broad unit of inheritance – The whole cellular architecture – including DNA, chromatin structure, etc. – is inherited (Jablonka and Lamb, 1995)	7) Narrow unit of inheritance – Only the nucleotide sequence is inherited

I will reflect more on the implications of these differences in the final chapter, but, with this table in front of us, it is easy to preview some of them. Issues of extended personal, social, and legal responsibility arise from points one and two (Rothstein, Cai, and Marchant, 2009; Dupras, Ravitsky, and Williams-Jones, 2014; Hedlund, 2012). Point three, which situates mothers and their behaviors at the center of epigenetic attention, suggests possible increasing intervention in the maternal body (Richardson, forthcoming; Richardson et al., 2014), which may be racially inflected (Mansfield, 2012; Mansfield and Guthman, 2015).

Point four speaks to the risk of classifying human groups because of the 'scars' left on them by challenging social conditions (Katz, 2013; Meloni, 2014; Meloni and Testa, 2014). And point five may inspire new privacy protections (Rothstein, Cai, and Marchant, 2009).

For now, let us explore more deeply how these differences between genetics and epigenetics provoke a new style of reasoning in biology. This novelty is particularly visible in the molecularization of the environment, in the complexification of the classical gene-environment interactionism, and in the non-genetic, developmental view of health and disease transmission across generations.

Postgenomic epistemology

Molecularizing the environment

The new thought-style emerging with postgenomics and epigenetics undermines the nature-nurture dichotomy on both sides. To the extent that genes are now "defined by their broader context" (Griffiths and Stotz, 2013: 228), our understanding of nature becomes inextricably entangled with social and environmental factors. Genes are socialized entities. As Meaney (2001: 52, 58) wrote more than a decade ago, "There are no genetic factors that can be studied independently of the environment, and there are no environmental factors that function independently of the genome." Moreover, because "environmental events occurring at a later stage of development...can alter a developmental trajectory," linear regression studies intended to discern the relative contributions of nature and nurture are meaningless.

This is not a one-way process of dissolution of nature into nurture. Much as genes are contextualized, context is molecularized. Environmental, social, and experiential factors there are translated into signals at the molecular level (Landecker, 2011, 2016). Social categories (race, class, social position), environmental factors (maternal care, nutrition, toxins), and bodily processes (metabolism) are being reconfigured today in molecular terms (Landecker, 2011; Niewöhner, 2011). As Landecker writes:

> If gene expression is hypothesized in epigenetics to be altered by environmental factors acting on genetic regulatory mechanisms then the structure of experiment must include particular practices and concepts that formalize "environment" as part of that system. (2011)

In other words, the environment is being "pinned down to the action and movements of particular molecules in the cell." The emerging discipline of nutritional epigenetics, for instance, focuses on "the way in which food affects patterns of gene regulation. It is a resolutely molecular science focused on how the molecules in food interact, via metabolic systems, with the molecules that attach to DNA and control levels of gene expression in the body (Landecker, 2011).

More generally, Landecker argues:

> In the contemporary moment, human social interactions, processes and material cultures are being investigated as biologically meaningful because they act as environmental signals that are epigenetically inscribed. Social things become biological things because they are transduced into the body as material patterns of chromatin conformation that persist as gene expression potentials, physiological manifestations or epigenetic memories. (Landecker, 2016)

Another way of looking at this reconfiguration of the environment is to think in terms of its digitization. As Meloni and Testa have argued:

> Epigenetics promises to capture the analogical vastness of environmental signals through the digital representation of their molecular responses. What seemed irreducibly analogic (the social, environmental, biographical, idiosyncratically human) needs to be overlaid onto the digital genome in a dyadic flow of reciprocal reactivity, and it seems this overlay can succeed only once the analogic is interrogated, parsed, and cast into genome-friendly, code-compatible digital representations (RNA, DNA associated with specific chromatin modifications as in chromatin immunoprecipitation, methylated DNAs, etc.). (2014)

Anthropologists and sociologists have been sensitive to this molecularization and miniaturization of the environment for the purposes of epigenetic research (Niewöhner, 2011; Landecker and Panofsky, 2013; Lock, 2012, 2015). This may produce a "ontological flattening" whereby "different categories of things in the world" – from motherly love to toxins, food to class inequalities – "will be made equivalent by recasting them as different forms of exposure" to a catchall phenomenon called the environment (Landecker and Panofsky, 2013; Meloni and Testa, 2014). This is a serious cause for concern, although the most sophisticated postgenomics researchers object that although nurture and the environment are made "mechanistic" by these molecular models, these are non-reductionistic views of mechanism, looking "upward to higher levels" (Bechtel, 2008: 21). This new version of mechanism, Griffiths and Stotz claim, is producing an unexpected rapprochement with themes from the holistic, or "integrationist," tradition (Griffiths and Stotz, 2013: 103).

Reconceptualizing interactionism

However, this molecularization of the environment is not the only discontinuity introduced by epigenetic research. While epigenetics is sometimes framed as an extension and an enrichment of the G×E interactionist paradigm (Meaney, 2010) – where genes and environment are distinct but interact – the rendering of the environment in postgenomics is very different and actually complicates or even undermines the G×E model (Charney, 2012; Landecker and Panofsky, 2013).

G×E models assume genotype and environment are neatly divided and their interactions must be tracked in order to analyze causes of phenotypic variation in a population. The model is typically represented by a diagram in which different allelic configurations modulate phenotypic susceptibility to a certain environmental input. The gene is pictured as an "invariant" entity, while the environment is "ongoing and non specific" (Landecker and Panofsky, 2013).

Epigenetic perspectives contrast with this model in three ways. First, they reverse the directional arrow. The question is not so much "how genetic variation modifies the sensitivity of the body to the environment" but rather "how environments come into the body and modulate the genome." The problem for epigenetic models is not which allele is expressed in which environment but "whether and to what degree a gene is transcribed and translated – and when and in what tissue.... Different individuals may have exactly the same DNA sequence, at any gene location, yet have different epialleles – different epigenetic modification and/or expression of that sequence" (Landecker and Panofsky, 2013).

Second, genetic functioning and environment are no longer beyond the effects of time. Epigenetics introduces

> a temporal logic that results in important consequences for study design. Most documented epigenetic effects result from an environmental exposure that occurs at a critical developmental moment; the work around aging and genome instability in cancer and neurodegenerative conditions also points to significant change over the life

> span with tissue heterogeneity over time. The effect of any exposure depends on when it occurs, and the effect might persist long after exposure has ceased. (Landecker and Panofsky, 2013; see Lappé and Landecker, 2015)

A third and related difference from G×E models is the inclusion of interacting non-genetic factors that would be inadequate to cover under the environment label: "heritable epigenetic variability (...) determined by epigenetic states that are environmentally induced and also possibly inherited from previous generations" (Tal, Kisdi, and Jablonka, 2010).

These three differences indicate that, with epigenetics, we are entering a messy, entangled terrain where interactionist discourses, based on bringing together separated sources of causality, are no longer accurate. Epigenetics introduces a new view of G×E interaction – G×E×HE (heritable epigenetic factors; see similarly Tal, Kisdi, and Jablonka, 2010). Even if one aims to keep only the variable E as source of HE, that variable is dramatically expanded thanks to the long-term temporal dimension epigenetics adds to environmental effects on genes. Epigenetics offers the possibility of transmitting G×E not only over someone's lifetime via cellular reproduction but also from one generation to another, even when the environmental signal has been switched off (Dulac, 2010). Epigenetics is therefore less concerned with the interplay of genes and environment in the present than the "inherited effects of the interplay of genes and environment" (Jablonka, 2004), "effects that last for many generations even when the environment no longer induces the phenotype" (Jablonka and Lamb, 2014). Epigenetics is plasticity across time and generations (Jablonka, personal communication). The difference between this paradigm and gene-centric ones becomes especially clear when considering developmental and epigenetic models of health and disease.

DOHaD: a non-genetic model for public health

Developmental origins of health and disease (DOHaD) is a research program at the intersection of "experimental, clinical, epidemiological and public health research" (Gluckman et al., 2010). DOHaD was created in the 2000s, but its origins go back to the 1980s when physician and epidemiologist David Barker started his pioneering work on the fetal origins of coronary heart disease. Barker, whose work was initially met with skepticism, established a link between risk of cardiovascular disease in adult life and fetal malnutrition (1995; see also Barker, 1998). Later he expanded the timescale to include early life nutrition.

DOHaD may be seen as a generalization of Barker's work on early developmental effects on susceptibility to a broad range of non-communicable adult diseases, from obesity to mental illness. The generalized form adds a potential mechanism for these delayed effects of prenatal and perinatal life into adulthood: epigenetics. Today DOHaD has become an internationally institutionalized reality with a professional society boating 550 members from 57 countries, biannual congresses, and a dedicated *Journal of Developmental Origins of Health and Disease*. Though established only in 2009, the journal has taken off, with 44,000 citations or references on Google and 10,500 on Google Scholar just in 2014 (Hanson, 2015)[4]. Though DOHaD has thus far had little public policy impact (except in Singapore and China, see HPB, 2010), it is becoming more prevalent in scientific discourse.

The main proponents of DOHaD, Mark Hanson and Sir Peter Gluckman, are adamant that this new paradigm has its rationale in the inability of gene-centric theories to explain non-communicable disease such as obesity and cardiovascular disease. As they claim, "Genes Aren't Us": "It was gradually acknowledged that the strong genetic determinism which had driven much of biomedical research for two decades was not particularly helpful in understanding the human condition" (Gluckman and Hanson, 2012: 96).

The theoretical alternative to gene centrism for the understanding of health and disease is the notion of developmental plasticity, the non-genetic adaptation that enables a phenotype to cope with changes in the environment:

> Far from passively adhering to a rigid genetic blueprint during early development, the developing organism is receptive to external cues and responds by adjusting its phenotypic development, thus resulting in several different possible phenotypes arising from a single genotype. It achieves this via the processes of developmental plasticity, which are ecologically prevalent and seen across many taxa. (Gluckman, Low and Hanson, 2013: 32)

Epigenetics is considered the molecular mechanism that underpins developmental plasticity; it does so by regulating "gene expression in response to environmental cues through covalent modification of DNA and its associated molecules" (ibid., see also Low et al., 2011). Equipped with this developmentalist view of biological processes, DOHaD investigators have linked metabolic disorders, cardiovascular disease, and mental illness to in utero and early life experiences of nutrition and behavioral stress.

It is important to clarify that this model is neither G×E interactionism nor a novel environmentalism determined to undo the "tyranny of the genes." Gluckman and Hanson claim that the key to adult health does not lie in the lifestyle of the current generation. The developmental trajectories determining (programming) susceptibility to disease (our "fate" or "paths of destiny" in DOHaD parlance, see Gluckman and Hanson, 2012) have been set for life, via epigenetic mechanisms, in the first post-natal years, our fetal environment, or even earlier, during the life of our parents or ancestors. Basically, it is the mismatch between these early settings ("the induced phenotype") and conditions encountered in adult life that create disease susceptibility in human populations (Gluckman et al., 2010; see also Gluckman and Hanson, 2006 where however the emphasis was more on a standard narrative of a mismatch between our evolutionarily evolved nature and current life conditions).

This is also why a new regime of prognostication and treatment based on epigenetic biomarkers taken at birth and followed by "lifestyle interventions in early life," when the epigenome is "highly sensitive to environmental factors," is increasingly considered the best way to assess and control future risks (Godfrey et al., 2011; Gluckman and Hanson, 2012; Hanson et al., 2011). This is far from environmentalism, as we shall discuss in Chapter 8. But this is not trivial interactionism either:

> Gene-environment interactions have traditionally been used to explain phenotypic variation in humans, such as differential susceptibility to disease. However, such a dichotomous perspective is increasingly becoming outmoded in light of the accumulating data pointing to the role of early life development in disease causation. (Gluckman et al., 2010)

The key is not in genes or in lifestyle but in developmental events that have programmed our physiology (for a critique of programming, see Waggoner and Uller, 2015). DOHaD aims to understand "how the processes of developmental plasticity – whereby phenotypic development is influenced by developmental experiences – affect an individual's interaction with its environment at maturity and, in turn, its risk of metabolic disease" (Gluckman, Low and Hanson, 2013: 32).

DOHaD can be considered a significant endpoint to our century-long story of the hard-versus-soft-heredity debate. Between epigenetics and DOHaD we have a fully developmental view of heredity that conjoins the two poles separated since 1900 by the rise of hard heredity. We saw how hard-hereditarian views emancipated themselves from developmentalist and generative views of heredity in which parents' or grandparents' experience had a direct formative influence on offspring phenotype. The key move of hard heredity was to abstract heredity from relations between parents and offspring. With DOHaD we return to developmentalist views where the lifestyle of our ancestors becomes the key factor in disease risk. On this reading, what happens in the womb, doesn't stay in the womb. However, what happens in the womb may be itself the product of what happened to the mother before pregnancy (for instance her nutrition and lifestyle well before pregnancy), at the time of her birth, and even in the lives of her parents. To remake another idiom, we are not so much what we eat but what our parents ate.

The epigenome is historical memory: the molecular archive of past environmental conditions (Heijmans, Tobi, and Lumey, 2009). Our ancestors' experiences "manufacture" our biological features; their lifetime and ours is united. The new heredity, like the old developmentalist one, does not end at birth. Thus Hanson and Gluckman have not shied away from talk of soft inheritance being back, alive and well. "Soft inheritance has now been reborn: the demonstration of developmental epigenetic processes provides a solid molecular basis for understanding how environmental influences can affect the phenotype of the next generation, or even those which follow, including susceptibility to chronic disease" (Hanson, Low and Gluckman, 2011).

Are we at the end of a150-year parabola, the making and the unmaking of hard heredity? Is heredity today coinciding with that blurred, confused nineteenth-century mechanism that, according to Charles Rosenberg, "began with conception and extended through weaning" (Rosenberg, 1974)? And if "a new softer synthesis" is replacing the old hard-hereditarian one, does this really "give scope for optimism" as its proponents claim (Hanson, Low and Gluckman, 2011: 11)?

8
Conclusions: The Quandary of Political Biology in the Twenty-First Century

In a sense, this book should have ended with the last question mark of the previous chapter. The transition to a postgenomic order is in its infancy. It is complex and unpredictable and already overhyped. Huge commercial interests are at stake[1]. Deterministic and holistic research strategies are in conflict (Morange, 2006; Richardson and Stevens, 2015; Waggoner and Uller, 2015). Any forecast of its future may be seen as insolent.

That said, I do not want to shy away from the intellectual provocations I have already made. Something is changing, and we should at least consider possible future directions, if not attempt firm predictions (Italians have an expression for this for which there is no standard English translation: *Lanciare il sasso e nascondere la mano*, literally: throwing a stone and hiding your hand, i.e., launching a provocation and pretending you haven't).

We are entering uncharted waters. If the hegemony of hard heredity is ending, at least in the rigid version we have known it at the peak of genecentrism, what will be the social ramifications of what, for simplicity, we have called a soft-hereditarian worldview? History is our only guide here, and it tells us that soft heredity cuts both ways. As we explored especially in Chapter 4, until hard heredity became dominant in the 1930s, soft-hereditarian views circulated in public health in two overarching forms: degenerationist and regenerationist.

Right-wing Lamarckians emphasized the degenerative effects of pathogenic environments on human germplasm. Alcohol, sexual diseases, and the moral and physical squalor of the slums all could poison heredity. Throughout the nineteenth and early twentieth centuries, doctors, educators, and social reformers were obsessed with the transgenerational perpetuation of these toxic environments and bad habits within poor families and groups deemed dangerous. Supposedly, indelible scars were left on the germplasm not only of the exposed but also of the unexposed generations. Some degenerationist thinkers even called on the state to strip poisoned groups of citizenship.

In opposition to degenerationism, left Lamarckians such as Paul Kammerer claimed that because heredity was "soft wax in our hands" it was possible to make it "comply with our wishes" (1920; quoted in Gliboff, 2005) by means of various techniques, many of which would be objectionable today. In a spirit that resonates with current hype about changing our genes through diet or exercise, newspapers of the 1920s celebrated Kammerer, the man who had regrown "eyes in sightless animals" and predicted a coming "race of supermen."

I don't mean to overplay these historical analogies. 2016 is not 1920, and epigenetics will not likely lead us back to medical degenerationism or Kammerer's regenerative eugenics as such. However, there is a sense in which these parallel politicizations of soft heredity – one focused on environmental scarring of the germplasm, the other on the reversal of these scars – have been reactivated today.

The epigenetic version of the first line of thought is not truly degenerationist, because it is mostly moved by a compassionate wish to show that historical and psychological traumas are "real," as they leave epigenetic marks on present and future generations. From the Dutch Hunger Winter of 1944 to 9/11, from the effects of bad parenting to the Holocaust, events of the near and more distant past, occurring in the family or in the broader society, are becoming sources of claimed biological damage.

Similarly, the contemporary version of the second line of thought is not truly regenerationist because Kammerer's framework was mostly utopian, collective, and socialist, whereas today's advice to take care of one's epigenome is fully neoliberal: epigenetic claims of regeneration are mostly mobilized to encourage individual techniques of care such as physical exercise, proper eating, smoking cessation, and, more mystically, meditation.

Nonetheless, there is a sense in which the old debate is being reactivated under new circumstances, which was literally unthinkable within a hard-heredity framework. Thus today we have magazine and newspaper headlines about exercise activating or deactivating gene expression. The stories underneath extol a cheap and accessible way to health and improved quality of life (Reynolds, 2014). On the degenerationist side, we have proclamations of "poisoned" heredity, as in a 2013 *Economist* report on negative transgenerational effects of folate deficiency in mice (2013). The double-edged sword of biological plasticity is as sharp as ever: Since bad experiences can turn into bad biology, is epigenetics bad news? Or is it good news because we can reverse the legacies of traumatic experiences? And who will be the "somatic experts" (Rose, 2007) who can reverse them? Doctors, social reformers, developers of epigenetic drugs?

But epigenetics is not only reawakening old questions and claims, under the new hyped and technologically driven biomedicine of the twenty-first century, in Western Europe or North America. Outside the boundaries of Western science as well, research traditions defeated in the course of the century of the gene are returning. Latin America, with its tradition of soft-hereditarian eugenics, is witnessing interesting appropriation in medical discourses of epigenetic arguments. But the spectacular case is Russia. Here historians are impressed by the impact of epigenetics in revitalizing nationalist and Lysenkoist views of science (Gordin 2015; Graham, 2015), and the multifarious political usages of the epigenetic discourse. Some argue patriotically that Soviet science (i.e. Lysenko) was right and present political power seems keen to capitalize on these nationalistic claims. On the other side, epigenetics' uses are always very flexible. A leading Russian immunologist has challenged Putin's power by suggesting that his countrymen are passive and obedient thanks to, in Loren Graham's words, "fears inherited from ancestors who endured the Stalinist repressions" (Graham, 2015). This claim referred explicitly to epigenetics and in particular a study on inheritance of olfactory fear conditioning in mice (Dias and Ressler, 2014)[2].

These are not yesterday's anxieties of degeneration, but they do emphasize damage passed from generation to generation with deleterious effects on the nation. And it's not only scientists looking to epigenetics to explain large-scale social distress. Something even more interesting seems to be happening at the grassroots level. Rodologia (from *rodstvo*, meaning kinship) is an amateurish self-help method based on the idea that various collective (gulags, war, Stalinism) and family (divorce, poverty, famine) traumas affect present well-being (Leykin, 2015). Scars of the past are imprinted on family genes and become manifest in later generations conditioning their self-realization. From reluctance to have children, to fear of divorce, people following Rodologian treatment trace their family history and genealogy not in order to study their genes but "to rearrange personal and familial narratives" of trauma (ibid).

Rodologia cannot be framed as a degenerationist narrative, because these scars are "both the origin of one's personal problems and the resource with which to solve them, participants are asked to turn to what they call the power of one's kin" (ibid). Still, *Rodologia* is a powerful example of the many uses of a return to soft-hereditarian claims.

Many of these claims are speculative, and some are pseudo-scientific. In human epigenetics, we have a mixed bags of findings corroborated by solid epidemiological studies and more speculative visions. Something similar occurs also for more strictly social science applications of epigenetic and in general developmentalist approaches. Three sites of sociological investigation are feeling these effects: race, gender, and class.

Race in epigenetic times

In 2009 the *American Journal of Human Biology* published a study looking at birth weight disparity between African and European Americans as a biological consequence of slavery. The article is a theoretical contribution to an ongoing attempt to crack an "epidemiological enigma" (Collins et al., 2011): the persistent black-white gap in mean birth weight. African American women are "twice as likely as European American women to deliver children with weight below 2.5 kg" (Jasienska, 2009).

Grazyna Jasienska, an ecologist with expertise in women's reproduction, points out correctly that approximately two-thirds of this difference does not have a genetic explanation. However, even discounting present environmental factors such as socio-economic conditions, nutrition, and prenatal care, "significant racial difference in birth weight still remains." This is where the slavery hypothesis[3] kicks in:

> Since neither genetic factors nor current socioeconomic determinants can adequately explain the existence of birth weight variation between 'races,' other environmental, social, and historical reasons must be considered. The following observation points to slavery as the factor of potentially profound importance: contemporary black women who were born in African countries ancestral to slave populations, but who live in the United States, give birth to children with significantly higher weight than black women in the United States who have slave ancestry. (16)

Though medical data on slavery are uncertain and disputed by historians, and though birth weight is a complex evidentiary terrain, Jasienska attributes, to "a large extent," the low birth weight of contemporary

African Americans to "conditions experienced during the period of slavery."

Low birth weight at the time of enslavement seems plausible: maternal undernutrition and intense physical labor, and arrested childhood development in mothers-to-be under slavery, were obviously potential causes for smaller babies, given the trade-off mothers faced between their own and a growing fetus's physical needs (ibid.). Moreover, in extremely stressful contexts, smaller babies have the adaptive advantage of reducing the energy demands mothers face during offpsring's infancy and childhood (Wells, 2010). But what is the causal link between slavery of years past and current low birth weight issues among African Americans?

Jasienska recognizes that the specific physiological mechanisms of this long-term programming "are unknown at present" or "not well understood" . Epigenetics is flagged, in passing, as the possible pathway by which "intergenerational information about environmental quality can be passed to next generations."

Although Jasienska may appear on shaky ground, her speculation is worth taking seriously for at least two reasons. First, evolutionary studies frequently assume that "birth weight in contemporary populations is not only determined just by the current maternal condition, but also by the influence of intergenerational life conditions, i.e., influences integrated across several generations" (16; see Kuzawa, 2005; Wells, 2007a and b). These studies claim that the developing fetus tracks not only present nutritional information but also, as a safety measure in a too-rapidly changing environment, "information about past environmental conditions that serves as cues for making predictions about future environmental conditions" (Jasienska, 2010). A baby's birth weight is determined not only by maternal and grandmaternal birth weight "but also very likely by its great-grandmother's birth weight and her childhood nutritional status" (Jasienska, 2009: 17). "Averaging across generations" makes the prediction about future environments more reliable (Kuzawa, 2005).

Second, the article's claims are justified inferentially by analogy to a more recognized example of measurable multigenerational effects on birth weight: the previously examined Dutch Hunger Winter. It is mostly by reference to the Dutch case, where "women suffering from famine during the last trimester had babies with almost 300 g reduced birth weight in comparison with babies born after the famine," that the slavery argument makes sense. If epidemiological studies on the Dutch famine have shown so consistently the multigenerational impact of a short-term case of malnutrition, how much deeper should this effect be for African Americans who "suffered much longer-lasting nutritional deprivation" and extreme energy expenditure? (Jasienska, 2009: 21, 22)?

Jasienska's study is neither the most famous nor the most cited among new epigenetic views of race. A better-known approach to racial health disparities is Christopher Kuzawa and Elizabeth Sweet's article, "Epigenetics and the Embodiment of Race" (2009). The key argument is that the cumulative effects of chronic social stress produced by contemporary racism (not nineteenth-century slavery) and systematic discrimination negatively affect maternal biology and are transmitted at each successive generation via the intrauterine environment, programming the fetus for a higher risk of cardiovascular disease. Racism goes under the skin and becomes literally the biology of future generations, producing and reproducing biological differences.

Both articles are powerful reconceptualizations of race in epigenetic terms (Meloni, 2016). Both reactivate, by means of twenty-first-century science, old ideas of race as the accumulated memories of past biosocial experiences. Renewed interest in the exquisitely Lamarckian notion of "biological memories" (Thayer and Kuzawa, 2011) or "phenotypic memory" (Kuzawa and Quinn, 2009) speaks to the revivification of forgotten or marginal traditions of thought in the epigenetic present.

However, there are several differences between the two articles. Kuzawa and Sweet's article points much more to the present perpetuation of discriminatory effects in American society as the basis for the reproduction of multigenerational epigenetic effects. While the social consequences of race can have "durable" effects on biology and health, this durability does not equate with permanence; "epigenetic effects are not set in stone, and may be amenable to reversal by intervention" (ibid.).

Jasienska's article, on the contrary, emphasizes the persistence of past traumatic events. The author is adamant on the fact that the effects of slavery may be "recalcitrant" or "resistant" to present changes and not easily amenable to social policy reform:

> Several generations that have passed since the abolition of slavery in the United States (1865) has [*sic*] not been enough to obliterate the impact of slavery on the current biological and health condition of the African-American population.... Even when the mother is well nourished herself, an intergenerational experience, which may be integrated in her own maternal physiology and anatomy, may cause her organism to follow the physiological strategy, which results in a reduced birth weight of her children. (16, 22)[4]

Jasienska concludes, "It is hard to predict how many generations with improved nutrition and health condition are needed for a significant increase in birth weight" (21)[5]. The resilience of acquired biological features is low and sometimes null. As she writes, "Too few generations have elapsed with improved energetic status to counteract the tragic multigenerational effects of nutritional deprivation on birth weight of children." This argument is as close to degenerationism as one finds in contemporary scientific study. It is not quite degenerationism, of course, but it does concern a biological abnormality reproduced in certain human populations as a consequence of historical events and biological plasticity (Mansfield and Guthman, 2015)[6].

It is interesting to compare this line of thinking to a different counterpoint, Boas's studies on human plasticity in immigrants and their descendants in the United States. Boas recognized that environmental influences could alter even the supposedly unchangeable cephalic index (1910, and 1912), and his work is still considered "the first authoritative statement on the nature of human biological plasticity" (Gravlee et al., 2003; a more sceptical view in Sparks and Jantz, 2003). His view was, in our terminology, an epitome of regenerationism: he looked, as Kammerer did, to race as an always-unstable category ready to be made and remade by changing historical conditions. The American soil was a great equalizer. Eastern European Jewish or Southern Italian immigrants arrived in the country long-headed or broad-headed, but in just a few years their children's measurements converged: "not even those characteristics of a race which have proved to be most permanent in their old home remain the same under our new surroundings; and we are compelled to conclude that when these features of the body change, the whole bodily and mental make-up of the immigrants may change" (Boas, 1910; see Degler, 1989; Gravlee et al., 2003).

Jasienska's study instead is backward-looking. It focuses more on the difficulty of changing the present in light of the past, an inertial phenomenon, to borrow from Kuzawa, in which the physiology of each generation is less shaped by the present environment than by an average of present and past cues "sampled over decades and generations" (Kuzawa, 2005: 12–13). Plasticity does not respond to "current ecological signals, but to parental cues, which tend to integrate past environmental experience." This "calibrates offspring biology to something akin to a running average of conditions experienced in the recent past" (Kuzawa and Thayer, 2011: 35). Kuzawa calls conditions experienced by recent matrilineal ancestors a "best guess" of "conditions likely to be experienced in the future" (Kuzawa and Bragg, 2011).

Clearly, two philosophies of plasticity are confronting each other. This tendency of epigenetics to become a science of disruptive factors rather than regenerative ones is challenging. It finds expression in epigenetic research concerning not only race, but also gender and class.

Gender in epigenetics

In expanding our understanding of inheritance to include ancestral environment and lifestyle, epigenetics extends responsibility across generations (Hedlund, 2012). But this extended responsibility appears often strongly gendered. The extreme attention to the role of mothers in DOHaD and similar lines of research has become the focus of criticism (Mansfield, 2012; Richardson, 2015; Richardson et al., 2014). The new research seems to double down on concerns about lifestyle and health of women before pregnancy, treated as if they were "eternally pre-pregnant" (Waggoner, 2015). The concern is that DOHaD epigenetics more generally augur further control of the maternal body.

Writing about DOHaD, Sarah Richardson (2015) notices insightfully that in epigenetic experiments, modification is invariably "introduced via the behavior or physiology of the mother." In most DOHaD and epigenetic research, Richardson highlights, it is the maternal body that becomes "an intensified space for the introduction of epigenetic perturbations in development." Women are "the central targets of health intervention" (2015; see also Richardson et al., 2014; Mansfield and Guthman, 2015). She refers to the agouti mouse study and Meaney's rat experiments discussed in Chapter 7, and to a study on changes in vole coat thickness via maternal melatonin.

Indeed, Kuzawa emphasizes stress and physiology in African American mothers as the key channel through which intergenerational effects of racism are reproduced and perpetuated. "Although less often studied, research suggests a more limited but biologically important lingering impact of paternal stress experience on the biological characteristics of offspring," Kuzawa writes in passing. Jasienska's study centers on the role of slave mothers; "intergenerational effects seem to be stronger for female lines than for male lines" (Jasienska, 2009: 17). The biologist Jonathan C. K. Wells has developed notions of "maternal capital" and the "metabolic ghetto" (see for instance, 2007 and 2010) to highlight the link between maternal biology and offspring's susceptibility to disease. In his challenging theoretical framework, maternal and grandmaternal phenotype shape the biological features of future generations, while paternal effects are largely invisible (Wells, 2010).

DOHaD-driven advice mostly concerns the lifestyle and health of women – from their earliest days – as the critical step toward "health in the next generation" (Wells, in Richardson, 2015). In this context, the fetus, "as a crucial node in space-time, simultaneously archiving the past while becoming the future," (Mansfield and Guthman, 2015) is the real concern, and the well-being of the mother is merely instrumental (Richardson, 2015). Pregnancy itself is considered "a niche occupied by the fetus" (Wells, 2007a and b, 2010). Mothers are advised to eat well, breastfeed, quitting smoking and other damaging habits (Gluckman and Hanson, 2012). They are exhorted "to make lifestyle changes in the service of their genetic lineage, while maintaining that these changes are unlikely to bring them or their offspring any benefit" (Richardson, 2015).

As other critics have noticed, this intervention in the maternal body is not only obsessive but more seriously unrealistic. Arline Geronimus points out that DOHaD literature considers women "gestating mothers with interchangeable composites of placentas, amniotic fluid, chemical reactions and molecules." This is a caricature of real life. DOHaD also promotes "maternal behaviors that are inconsistent with women's social realities.... For example, as sociologists, family demographers, and ethnographers have observed, breastfeeding is at the very least impractical to expect of many poor mothers without other social changes" (Geronimus, 2013).

This social and gender critique of DOHaD is appropriate, timely, and insightful, but does it apply to epigenetics research in general? Oliver Rando at UMass argues that epigenetics offers methods for discerning how "paternal environmental conditions influence the phenotype of progeny" (2012). Indeed, germline inheritance of paternal effects of growing importance in epigenetic research (Curley, Mashoodh, and Champagne, 2011; Rando, 2012; see also Puri, Dhawan & Mishra (2010), and some of the most challenging discoveries in epigenetics come from the paternal line. One obvious case is the Överkalix cohort work, which emphasizes male-line transgenerational effects of food availability on longevity (Kaati et al., 2002; Pembrey et al., 2005). Perhaps the DOHaD approach is just not representative of epigenetics generally. Or, more profoundly, a sexist society has hijacked scientific research for its own goals, sidelining alternative approaches that might foster greater equality. This sort of hijacking seems especially prominent in epigenetic approaches to class.

Epigenetics and class: pathologizing the poor?

If epidemiological studies use epigenetics to record disruptive or adverse effects of race and gender (Richardson, 2015; Mansfield and Guthman,

2015), they also do with respect to class. Although in principle epigenetic studies of social position need not focus on absolute conditions of poverty – any change upward or downward could be studied (Niewöhner, 2011) – this neutral stance seems to have been lost. Anthropologist Jörg Niewöhner, who in 2007 did one of the first ethnographic studies of an epigenetics lab, noticed that the initial design of a methylation study to measure social position was an impartial one, which procured impartial results:

> Initial and as yet unpublished findings indicate that methylation status at a number of sites changes more within subjects that have experienced a change in their socio-economic status from birth to their 40th birthday compared to subjects that retain the same status – even if that is a low status. Thus, epigenetic modification may be more sensitive to relative change than to a low socio-economic status in absolute terms. (2011)

However, epidemiological uses of epigenetics have since gone in another direction. Most studies have concerned trauma, stress, stressors, deprivation, toxic exposures, social insults, adversities, scars, wounds, famine and malnutrition, poor parenting, poverty, separation, and early adversity in absolute terms. The environment under consideration in epigenetic studies is one of unchanging adversity, if not calamity. The prevailing research design looks at the link between social deprivation and abnormal or hypomethylation; epigenetics therefore becomes the signature of poverty.

A 2012 study by Dagmara McGuinness and colleagues exemplifies this trend in epidemiological epigenetics[7]. McGuinness studied associations between socioeconomic and epigenetic difference in Glasgow, a city known for its "extreme socio-economic gradient of health inequality... which is not fully explained by conventional risk factors for disease." The difference in life expectancy between the richest and poorest areas of the city is enormous: nearly thirty years in just a few miles' distance (Marmot, 2005; Reid, 2011). And Glasgow's poor seem to be much worse off those of similarly depressed areas of deindustrialized U.K. cities such as Liverpool and Manchester. Epidemiologists call this the Glasgow effect (Reid, 2011).

To understand this specificity, the study investigated methylation levels obtained from blood samples of 239 members of the pSoBid – Psychological, social and biological determinants of ill health – cohort, which gathered people from the richest and poorest areas of Glasgow.

The researchers found that the poorest and those with the lowest in job status (manual workers) had lower DNA methylation than the rest. DNA hypomethylation was not associated with age but with class, to the point that years of education – which in the United Kingdom is strictly related to class – correlate positively with global DNA methylation (McGuinness et al., 2011).

The significance of these findings can of course be questioned. *Prima facie*, it is not surprising that hypomethylation, which is associated with risk of inflammation (e.g. immune responses), is more pronounced in groups exposed to greater social hardship. But once again the framework for this study is not merely environmentalist, i.e., hardship or poor nutrition have a direct effect on methylation. A developmentalist logic is introduced with the claim that one of the reasons for global hypomethylation could be developmental programming *in utero*[8].

The study's circulation in broader society is even more interesting. Unsurprisingly it was the programming effects of methylation *in utero* that took hold of the popular imagination. *The Irish Times* spoke of "cards of life...dealt just weeks after conception, when methylation takes place in the embryo." "Babies born into poverty are damaged forever before birth," *The Scotsman* declared, "the health of babies born in deprived areas could be damaged for the rest of their lives long before they have even left the womb" (see respectively, Hennessy, 2012; McLaughlin, 2012). The papers were not alone in using crude deterministic language. Paul Shiels, one of the investigators, offered the following metaphor: "If you think of your chromosomes as the hard drive of a computer, and the methylation as a program, sometimes the program can be corrupted. If you have a poor program, then it's not going to work as well." "There's a drip effect," he adds, "which predisposes you from birth to be less robust, therefore more prone to early-onset disease" (quoted in both *The Irish Times* and *The Scotsman*).

It is not obvious what should be done with these findings and this language. The choice depends on stakeholder interests. For charities and National Health Service operators, the study offered "startling evidence" of the damage that poverty does to children even before they are born and served to reaffirm a "commitment to tackling poverty in" Glasgow (*The Irish Times*). In one of the dialogues in their seminal *Evolution in Four Dimensions* (2014), Eva Jablonka and Marion Lamb argue similarly that epigenetic findings may justify more public provision for disadvantaged people:

> The persistence of ancestral epigenetic states means that methods of compensating for the misfortunes of ancestors may be needed

> to ensure that the present generation does not start with an epigenetic disadvantage. Merely ensuring that individuals who carry detrimental ancestral marks develop in a normal environment may not be enough; it may be necessary to provide people with special diets or other treatments that will counteract their epigenetic heritage.

Political theorists working in a normative framework have claimed similarly that "knowledge of epigenetic mechanisms may increase our ability to achieve…equality of opportunity, by unraveling the mechanism through which the health prospects of a population are affected by the unequal choices and circumstances of their parents" (Loi et al., 2013).

This all sounds desirable, but how likely is it in a society where class, race, and gender inequalities remain so vast? What is our society going to make of the notion that, to paraphrase Filipchenko's argument, the socially disadvantaged are also (epi)genetically damaged? The racial other, because of its history, is also a biological other? And what will oppressed groups do with this flurry of epigenetic studies concerning their own condition? Some may see hope in epigenetics, but others will likely share the sentiments expressed by one *Scotsman* reader in an online comment following the paper's report of the Glasgow study:

> I am just flabbergasted by this latest research – I am 81 years old and was born into what I would describe as extreme poverty…but with caring parents who were not into accepting "charity" but gave me and my siblings the best they could in spite of a lot of unemployment. I have led a useful life, was pretty intelligent at school, and held responsible jobs, have married successfully, had children…and feel I was anything but deprived or damaged. Just grateful that these statistics weren't available in my past!

Conclusion: heredity reinforced?

It is possible that epigenetics will reinforce discourses about the biological difference or inferiority of the "undeserving poor" (Katz, 2013), which has long worked as a mechanism of class formation. Hard heredity, as we have seen, provides an easy argument about faulty genes running in certain families or races. However, soft heredity can be used to support similar claims, though following a different line of argument, e.g., that poverty running in certain social groups damages (a twenty-first century equivalent of "poisons") their offspring before birth.

We didn't have to worry about soft hereditarian race and class formation until recently. But epigenetics has done away with the abstract universal body. We have in a sense returned to the idea of a specific biology characterizing local groups – groups that differ because of their different stories. Are blacks biologically different from whites? Rich from poor? Under an epigenetic paradigm, these explosive questions are no longer impossible. Moreover, there is an important temporal dynamic in epigenetic class formation: Why do the poor stay poor? Why don't they progress? Why does their poverty resist stubbornly and reproduce over time? Especially in an age of increasing inequality, political uses of epigenetics may ask if the poor suffer an ongoing accumulation of bad biology and whether this – as opposed to, for example, economic structures – is responsible for them slipping farther behind.

After 1945 eugenics was increasingly sidelined and a new post-eugenic conceptual repertoire was created in Europe and North America to address once again the issue of the multigenerational continuity of poverty. It was now possible to talk about the dysfunctional Negro family in a "culture of poverty," as in the so-called Moynihan Report (1965) in the United States or a "cycle of deprivation," proposed by the British politician Keith Joseph in early 1970s (Welshman, 2013).

Epigenetic explanations tap into these longstanding debates on the reproduction of poverty and may invigorate them by offering a visible mechanism for the biological perpetuation of race and class pathologies.

Without denying that epigenetics can be used in favor of liberal arguments (Heckman, 2012) and serve as a weapon against racism (Sullivan, 2013), classism or sexism, its underlying soft-inheritance view may be no less exclusionary than is a genetic view of social relationships. Developmentalist and epigenetic views reinforce the idea of a strong chain connecting generations, suffocating the implicitly emancipatory aspect of Weismannism: each generation may start anew. Even when compared to sociological or economic models of multigenerational effects, DOHaD and epigenetics' approaches seem to imply a bleaker view, even in their benevolent language. People are not so much stuck in place (Sharkey, 2014; see also Massey, 2013; Sharkey, 2008) but trapped in bodies, something that once again raises serious issues about the amenability to modification of this long-term legacy. What was before epigenetics a metaphorical multigenerational passage for instance of trauma (as in the case of psychoanalysis) has become today a material signature of something that literally goes on from parents to progeny. A few months ago, in the *American Journal of Psychiatry*, Rachel Yehuda, the most recognized authority in the field of epigenetic transmission of PTSD, published the first "conclusive study" to demonstrate altered methylation levels (the most significant epigenetic markers) in descendants of Holocaust survivors. "This is the first evidence in humans, Yehuda has commented, of an epigenetic mark in an offspring based on pre-conception exposure in a parent" (Glausiusz, 2014). As *Scientific American* quickly commented, "parents' traumatic experience may hamper their offspring's ability to bounce back from trauma" (2015).

We can't yet say whether epigenetics will fulfill its liberating potential or instead further racist or classist agendas. Epigenetics can help us rethink the relationship between the biological and the social world, and lots of emphasis has been put by researchers on the reversibility of negative epigenetic effects, but even the best conceptual framework is open to unpredictable sociopolitical outcomes. Today, as throughout history, scientific theories do not decide political values.

Notes

1 Political Biology and the Politics of Epistemology

1. Hard and soft heredity are much later, twentieth-century terms, coined when the controversy was already closed on the side of hard-hereditarian authors. Their capacity to describe accurately the complexity of the historical debate has been rightly questioned by historians (Müller-Wille and Rheinberger, 2012: 90). Darlington uses the notion of softness for heredity for the first time in 1959, as a synonym for inheritance of acquired characters. Genetics is the radical disproof of soft heredity, he claims, "for the genes are not soft, they do not blend" (1959: 53). Mayr defines soft heredity as "the belief in a gradual change of the genetic material itself, either by use and disuse, or by some internal progressive tendencies, or through the direct effect of the environment" (Mayr 1980). He also claims emphatically, "It was perhaps the greatest contribution of young science of genetics to show that soft inheritance does not exist" (ibid.: 17). See for a reconstruction and a critical viewpoint on this story, Gissis and Jablonka 2011.
2. See, respectively, Pembrey et al., 2006; Yehuda et al., 2005, 2014; Almond, 2006, 2011; Mazumder et al., 2010; Jasienska, 2009.
3. Epistemology is not used here, as in much Anglo-American philosophy of science, as a conceptual and normative analysis aiming to clarify what constitutes knowledge. My program is historically oriented, to reflect on the "conditions *under* which and the means *with* which things are made into objects of knowledge" (Rheinberger, 2010: 2–3). True, my analysis has normative implications, but my focus is on historically situated truth-regimes rather than ahistorical idealizations of how knowledge is supposed to be.
4. The story of the review was pretty troubled, as Jackson explains (2001): rejected by the *Saturday Review,* it appeared in print in the February 1963 issue of *Scientific American* (Dobzhansky, 1963).
5. My criticism is aimed at appropriations of the label biopolitics in social theory and political philosophy (see for instance Campbell and Sitze, 2013). At a more empirical level, an abundant literature uses the label "biopolitics" as a framework for situated research on global health, biomedicine, the politics of pharmaceuticals, security, foreigners, humanitarianism, race, and prison. These reside outside my critique about the abstractness and lack of historical in biopolitics. See for instance for an alternative theoretical framework centered on the notion of a politics of life (Fassin, 2007, 2009).
6. The strong continuity between Nazi and contemporary biopolitics is evident across Esposito's work: "Not only has the politics of life that Nazism tried in vain to export outside Germany – certainly in unrepeatable forms – been generalized to the entire world, but its specific immunitary (or, more precisely, its autoimmunitary) tonality has been as well.... The truth is that many simply believed that the collapse of Nazism would also drag the categories that had

characterized it into the inferno from which it had emerged. (Esposito, 2008a: 147–148) The point is even more explicit in his paper "Totalitarianism or Biopolitics," where Nazism is decisively defeated militarily and politically but less so culturally or linguistically. (Esposito, 2008b: 641). See for a critique Meloni, 2010.

7. On differences and analogies with Kuhn, see Harwood, 1986; Mossner, 2011; Jasanoff, 2012
8. Nazi "barbarous utopia" is an expression used by Burleigh and Wippermann, 1993.

2 Nineteenth Century: From Heredity to Hard Heredity

1. See Darwin's comment in *N notebook*: "Habits becoming hereditary form the instincts of animals. – almost identical with my theory" (quoted in Gissis and Jablonka, 2011) See also Darwin's 1842 sketch: "It must I think be admitted that habits whether congenital or acquired by practice [sometimes] often become inherited; instincts, influence, equally with structure, the preservation of animals; therefore selection must, with changing conditions tend to modify the inherited habits of animals. If this be admitted it will be found possible that many of the strangest instincts may be thus acquired." "Also habits of life develop certain parts. Disuse atrophies." (in F. Darwin, 1909: 18, and 1).
2. It has also to be remembered that Darwin looked at heredity with a set of concerns different from our twentieth century soft/hard heredity debate. In particular he was anxious to avoid the objection that blending inheritance (Jenkin's swamping argument, 1867) might destroy the accumulation of variations, making therefore natural selection ineffective (see Bowler, 1989: 62–63; Bulmer, 2004b).
3. This does not mean that there were no conceptual conflicts between geneticists such as Morgan and the architects of hard heredity, such as Weismann. But these conflicts were along a different axis – Morgan's experimentalism versus Weismann's speculation (see Allen, 1975; alternative reading in Maienschein, 1991; see Esposito, 2013), an accusation that also Johannsen launched to Weismann. On the difficult accommodation of Weismann into the Mendelian age, see Churchill, 2015.
4. As Johannsen wrote in 1911, when the modern biological conception of heredity was already achieved, "Biology has borrowed the terms "heredity" and "inheritance" from everyday language, in which they mean "transmission" of money, things, rights, duties, ideas and knowledge from one person to another or to some others – the "heirs" or "inheritors." The transmission of properties – these may be things owned or peculiar qualities – from parents to their children, or from more-or-less remote ancestors to their descendants, is the essential point in the discussion of heredity, in biology and law."
5. Burkhardt (1977) and Corsi (1988) have shown that Lamarck did not originate the idea of inheritance of acquired characters. The construct was

"ubiquitous" in Lamarck's time; "*The law of nature by which new individuals receive* all that has been acquired in organization during the lifetime of their parents is so true, so striking, so much attested by the facts," he wrote in 1815, "that there is no observer who has been unable to convince himself of its reality." As Corsi claims, "Although known as the first and major theorizer on the principle of the inheritance of acquired characters, Lamarck never expressed himself in such terms, and clearly – and rightly – considered himself as one of the many naturalists convinced that the development of an organ during the lifetime of an organism, or the appearance of however slightly behavioral propensities ('habits'), could be passed on to the next generation if a mating occurred between individuals who had experienced the same change. Yet, characters were not acquired or transmitted: only biological processes were. To Lamarck, organic fluid dynamics naturally gained strength within the parts more exposed to changing environmental circumstances, thereby contributing to their reinforcement and extremely gradual modification (....)" (Corsi 2011: 12)

6. It is fair to say that the term already appeared in adjectival form (eugenic) in 1833 in the *Oxford English Dictionary* with the meaning of "the production of fine offspring" (Turner, 2009; Richards, 2013).
7. Respectively by Alpheus S. Packard, 1885, see Bowler (1983), and by George J. Romanes, 1883, according to Noble (2011), or 1888, although the term neo-Darwinism appears for the first time (twice) in Butler (1880), but not defined in the sense of Romanes. For this latter neo-Darwinism was an hardening, courtesy of Weismann's germ plasm, of the original flexible Darwinism which made room for both natural selection and acquired characters. He claimed that (1895): "(...) the Neo-Darwinian school is in Europe seeking to out-Darwin Darwin by assigning an exclusive prerogative to natural selection in both kingdoms of animate nature". Already in 1885, American entomologist Alpheus S. Packard (here quoted from his later *Lamarck, The Founder of Evolution*) used instead the term "Neolamarckianism, or Lamarckism in its modern form" as a term to propose the belief that "many species, but more especially types of and families, have been produced by changes in the environment acting often with more or less rapidity on the organism, resulting at times in a new genus, or even a family type. Natural selection, acting through thousands, and sometimes millions, of generations of animals and plants, often operates too slowly; there are gaps which have been, so to speak, intentionally left by Nature."
8. There are several sources where Weismann pays homage to Galton, letters and published works (see Pearson, 1930). Of course, the resemblance between the views of Galton and Weismann does not mean identity. See for instance amongst many critics of this identification Bulmer 2004a: "Galton did not forestall the essential part of Weismann's theory, the partially-mistaken idea that the germ plasm of the zygote is doubled, with one part being reserved for the formation of the germ-cells." See also Churchill, 2015.
9. The initial anti-Darwinism of geneticists, and therefore of hard-hereditarians, has to be kept in mind when, in next chapter, we'll look at the critic of soft-hereditarians thinkers like Kammerer; although not all hard-hereditarians were anti-Darwinian, see for instance Pearson and biometricians, part of the conflict between Kammerer and geneticists is also a controversy around Darwinism

and evolutionism, as it was understood at the time. The German-Soviet movie Salamandra, 1928, well depict in its pro-Kammerer interpretation, this conflict where Mendelism is equated with anti-evolutionist themes (also by the decision in the movie to replace geneticists with priests and aristocrats).

10. Just said in passing, it is not by chance that today biology with its epigenetic and microbial complications, is eroding just this modernist and individualist view of biological processes (Dupré, 2012; Gilbert, Sapp, and Tauber, 2012)

3 Into the Wild: The Radical Ethos of Eugenics

1. 1900 can be taken as a symbolic *a quo* term. In America, the Station for Experimental Evolution at Cold Spring was created in 1904 and the Eugenics Record Office in 1910; the British Eugenics Society was launched in 1907, the same year as the first compulsory sterilization law in Indiana; the term *Rassenhygiene* was created in Germany a few years before, in 1895. Medical attempts at coerced sterilization precede 1900 going back at least a decade, see for instance Largent (2008) for the American context. Michigan had debated sterilization already in 1897, ten years before the Indiana law was passed, though the bill was voted down by legislators (ibid.).
2. Galton certainly alerted Darwin of the problem by writing to him that selection in the present society "seems to me to spoil and not to improve our breed" since "it is the classes of coarser organisation who seem on the whole the most favoured...and who survive to become the parents of the next [generation]." (quoted in Paul, 2006).
3. See his 1957 text where he claims that "the human species can, if it wishes, transcend itself", and that "a vast new world of uncharted possibilities awaits its Columbus".
4. The two persons most easily identified with Social Darwinism, Herbert Spencer and William Graham Sumner, died respectively, in 1903 and 1910, just when eugenics was powerfully emerging. Bagehot died much earlier, in 1877.

4 A Political Quadrant

1. The area of hard-hereditarianism both left and right-wing is obviously larger than the label Mendelism, given the presence and support for eugenics of biometricians (who were non-Mendelians and hard-hereditarians). One such example, Pearson, will be discussed later in the chapter in his opposition to Saleeby on alcohol policies.
2. As anticipated in my Introductory chapter on political epistemology, my analysis of the four positions, although embodied in specific historical figures and debates, has to be understood primarily at a structural level, in which different conceptual permutations (e.g., soft heredity) were made possible and therefore politically appropriable before the political-scientific controversy was closed.

3. Gissis (2002) highlights the association between Lamarckism and the social, rather than individual level, as a key to its "futurity" especially in the context of the French Third Republic.
4. See for instance in *The Dialectics of Nature* (1896) this passage: "On the other hand, modern natural science has extended the principle of the origin of all thought content from experience in a way that breaks down its old metaphysical limitation and formulation. By recognizing the inheritance of acquired characters, it extends the subject of experience from the individual to the genus; the single individual that must have experienced is no longer necessary, its individual experience can be replaced to a certain extent by the results of the experiences of a number of its ancestors. If, for instance, among us the mathematical axioms seem self-evident to every eight-year-old child, and in no need of proof from experience, this is solely the result of "accumulated inheritance." It would be difficult to teach them by a proof to a bushman or Australian negro."
5. My analysis can be connected to Stephen J. Gould's general point about the "betrayal of Lamarck" by self-styled neo-Lamarckians who not only "elevated one aspect of the mechanics-inheritance of acquired characters-to a central focus it never had for Lamarck himself" but also more importantly "abandoned Lamarck's cardinal idea that evolution is an active, creative response by organisms to their felt needs" (Gould, 1980). The right-wing neo-Lamarckians I am focusing on reverse Lamarckian optimism but retain the notion of a direct influence of the environment upon the organism. Some neo-Lamarckians addressed the dualism between activity and passivity in the relationship between environment and organism, between physiogenesis – "direct, involuntary response any organism can make when exposed to new conditions" – and kinetogenesis – "creation of structure by animal movement as for instance the giraffe's neck" (see Bowler, 1983: 62). On the historical and conceptual transformation of Lamarckism and the dissimilarity between Lamarck's doctrine and the many versions of Lamarckism, see also Gissis and Jablonka, 2011. As Gliboff points out in that collection, "After two hundred years, the theories that now pass as 'Lamarckism' would hardly be recognizable to its original author...or to its early supporters" (2011; see also Burkhardt 1977; Corsi 1998).
6. As McKee (1993) notices to illustrate the polysemy inherent in Lamarckian views, another sociologist, Robert Park, argued that freedom was "breaking down 'the instincts and habits of servitude' and slowly 'building up the instincts of freedom" (quoted in 1993: 62).
7. I am not claiming that Lamarckians monopolized degenerationist rhetoric. Right-wing hard-hereditarians had their own approach to racial degeneration. Lamarckians envisioned the pathogenic effects of the environment poisoning the germ plasm. Mendelians and biometricians saw the unfit were outbreeding the best stock in a population. True, for right-wing hard-hereditarians, the unfit were also the ones living in the most pathogenic environments, but this was simply an effect of selective processes. The worse stock failed in society and therefore was gathered in the worse areas.
8. The whole passage stated: "it is, by the way curious that the anti-Mendelians have not realized that Lamarckism would create even greater theoretical difficulties than Mendelism. If the effects of the environment are imprinted

on or assimilated by heredity, then centuries of poverty, ignorance, disease, and oppression should have ingrained a most undesirable heredity upon the vast majority of the human species, and engrained it so firmly that a few generations of improved conditions could not be expected to effect much amelioration. Mendelism, on the other hand, makes it clear that even after long-continued bad conditions, an enormous reserve of good genetic potentiality can still be ready to blossom into actuality as soon as improved conditions provide an opportunity" (Huxley, 1949: 187).

9. Wallace (1892) claimed that "the non-inheritance of the effects of training, of habits, and of general surroundings, whether these be good or bad, is by no means a hindrance to human progress, if, as seems not improbable, the results on the individual of our present social arrangements are, on the whole, evil. It may be fairly argued that the rich suffer, morally and intellectually, from these conditions quite as much as do the poor; and that the lives of idleness, of pleasure, of excitement, or of debauchery, which so many of the wealthy lead, is as soul-deadening and degrading in its effects as the sordid struggle for existence to which the bulk of the workers are condemned. It is, therefore, a relief to feel that all this evil and degradation will leave no permanent effects whenever a more rational and more elevating system of social organization is brought about."
10. Kevles coined the definition "reform eugenics" (1985) in the context of his analysis of Anglo-American Eugenics, without looking specifically at the Soviet movement, which may explain why he used that label – that is, why he understood this part of the quadrant merely as a correction of mainline eugenics.

    He included in reform eugenics second-generation eugenicists such as Blacker and Osborn, along with Haldane and Muller. His point was that these reformers were more prepared than mainline eugenicists to recognize the many unknowns about genetics, thus destabilizing some of the previous dogmas of the movement. "The reformers recognized, however, that hardly anything was known about precisely what role heredity played in the achievement, or lack of it, of the bulk of the population" (1985: 173). If this is the criterion, though, I do not see how Muller can be considered a reform eugenicist. He was not skepticical about what we can or cannot know and therefore apply to society.

    If by "reform eugenics" we mean a position by which people "were aware that man as yet knew too little about human heredity to enact sweeping eugenic changes, let alone usher in a eutelegenetic utopia," (1985: 193) the label is apt to describe the work of Carr-Saunders, Hogben, Penrose, but not Muller. Haldane is also problematic under this rubric since his radicalism and utopianism (see Daedalus, 1924) remained largely unconnected to any real movement in society.
11. Lamarckism, Muller claimed, "was racist because a poor environment would have to impress its deleterious effects on the progeny, resulting in a genuine inferiority of the downtrodden, the peasants, and the proletariat as well as invidious distinctions between the heredities of wealthy and impoverished nations."
12. I have treated more extensively this theme in my Meloni, 2016.

5 Time for a Repositioning: Political Biology after 1945

1. As Levy and Sznaider (2004) write:

    Before 1945, international law mainly regulated relations between states, confirming the parameters of the Westphalian order, whereas after 1945 the knowledge of the enormity of the Holocaust has come to provide the main impetus for the privileged position that human rights regimes currently enjoy in the international arena. [....] Two major UN conventions, one declaring Universal Human Rights as a new standard [10 December, 1948], the other declaring genocide an international crime [9 December, 1948], make up the foundation of human rights regimes. Formed in the immediate post-war period in a brief window of opportunity before the Cold War overshadowed most international arrangements, they were specifically designed to prevent another Holocaust and another Nazi party.

2. This of course does not deny that that Nazi crimes were used politically as a source of a novel post-war unity, especially functional to build an "international image" of anti-racism and obscuring national racist issues for the U.S. (Hazard, 2012) and the other winning nations (Selcer, 2012; see also Dudziak, 2000 and Furedi, 1998).

6 Four Pillars of Democratic Biology

1. Major sociobiologists such as Dawkins and Wilson flirted with eugenics; in 1978 Wilson proposed a future of "democratically contrived eugenics", but this remained very much in the background of the core philosophy of the movement.

7 Welcome to Postgenomics: Reactive Genomes, Epigenetics, and the Rebirth of Soft Heredity

1. Parts of this chapter have previously appeared as "The social brain meets the reactive genome: Neuroscience, epigenetics and the new social biology", *Frontiers in Human Neuroscience*, 8, pp. 309 (2014), and "Epigenetics for the social sciences: justice, embodiment, and inheritance in the postgenomic age", *New Genetics and Society*, 34(2) (2015), pp. 125–151.
2. Often, a technical distinction is introduced between intergenerational (or parental) and transgenerational effects, with the former shorter and limited to two generations, and the second spanning over multiple generations (see Grossniklaus et al., 2013; Heard and Martienssen, 2014).
3. Barker (1986) and before him Forsdahl (1977, 1978) have been amongst the first researchers to look at long-term effects of prenatal exposures. Thanks to Michelle Kelly-Irving for pointing to me the study by Forsdahl.
4. Amongst the various disciplines that have reacted positively to the DOHaD approach, economics is probably the most significant, see Almond, 2006; Almond and Curry, 2011.

8 Conclusions: The Quandary of Political Biology in the Twenty-First Century

1. According to a recent study (marketsandmarkets.com, Sept 2014) the rising epigenetic pharmaceutical industry will be a global market worth $8 billion by 2017, twice the value of 2014.
2. In passing, it is interesting to notice that, differently from the Lysenko's era, the reinvigoration of soft-inheritance in Russia today takes legitimacy from knowledge produced in Western science, not from an inward-looking reclusion in patriotic science: It is a patriotism constructed via global flows of knowledge.
3. It has to be noted that this is not the first time that a scientific hypothesis about the effects of slavery has been recruited to explain racial health disparities. The so-called "Slavery Hypothesis" was firstly advanced as a genetics (i.e. non-epigenetic) theory to explain present U.S. black/white disparities in hypertension (Wilson and Grim, 1991).The theory, which received wide criticisms by historians and social scientists (Curtin, 1992; Kaufman and Hall, 2003), maintained that a selective process took place during slave transport that favored "individuals with an enhanced genetic-based ability to conserve salt" (Wilson and Grim, 1991). It was this group that survived the brutal effects of the Middle Passage in which the major causes of death were "salt-depletive diseases such as diarrhea, fevers, and vomiting" (ibid.). This would explain the "greater frequency of individuals with an enhanced genetic-based ability to conserve salt" amongst African Americans (ibid.).
4. An obvious critique to this is that African American slavery is far from being ancient history, given how conditions of slavery continued in the South for decades after the war, and famine was endemic during Jim Crow; see for instance (Blackmon, 2009). I thank Simon Waxman for pointing this out to me. However, here my point is not on the historical truthfulness of Jasienska's hypothesis but on her argumentative style and transgenerational explanatory framework. For a circulation of the slavery hypothesis in epigenetic terms see Maxmen, 2012.
5. In a different study testing the validity of intergenerational effects on overgrowth and stunting on Maya children in Mexico we find a comment made in passing that "it may take three or more generations of development-low socioeconomic status under good health conditions to completely override the past history of poor health" (Varela et al., 2009).
6. See also from an economic viewpoint informed by DOHaD framework the idea that, given how "early-life health measures of blacks have stagnated since the late 1990s," with "a black infant currently more than twice as likely to die before age 1 as a white infant" this results in the fact that "a future of racial inequality is being programmed" (Almond, 2006).
7. Further similar works include: Borghol et al., 2012; Tehranifar et al., Stringhini et al., for a note of caution on the methodology see Heijmans and Mill, 2012.
8. As the authors claim: "The extent of DNA hypomethylation in the most deprived group of participants is intriguing. Such global hypomethylation

could be reflective of environmental exposures and/or diet during life, or a direct consequence of developmental programming *in utero*, or a combination of both. Notably, adjusting for diet does not weaken associations with global methylation status, suggesting that *in utero* programming or environmental factors would be causative for the observed global hypomethylation associated with lower SES" (2011).

Bibliography

Abu El-Haj, N. (2007). The genetic reinscription of race. *Annual Review of Anthropology*, 36(1), pp. 283–300.

Adams, Mark B. (1989). The politics of human heredity in the USSR, 1920–1940. *Genome*, 31(2), pp. 879–884.

Adams, Mark B. ed. (1990). *The Wellborn Science: Eugenics in Germany, France, Brazil, and Russia*. New York: Oxford University Press.

Adams, Mark B. (1990a). Eugenics in Russia, 1900–1940. In: Adams, M., ed. *The Wellborn Science: Eugenics in Germany, France, Brazil, and Russia*. New York: Oxford University Press, pp. 153–216.

Adams, Mark B. (1990b). Toward a comparative history of eugenics. In: Adams, M. ed. *The Wellborn science: eugenics in Germany, France, Brazil, and Russia*. New York: Oxford University Press, pp. 217–229.

Adams, Mark B. ed. (1994). *The Evolution of Theodosius Dobzhansky: Essays on His Life and Thought in Russia and America*. Princeton, NJ: Princeton University Press.

Adams, Mark B. (1994). Theodosius Dobzhansky in Russia and America. In: Adams, M., ed. *The Evolution of Theodosius Dobzhansky: Essays on His Life and Thought in Russia and America*. Princeton, NJ: Princeton University Press, pp. 3–11.

Adams, Mark B. (2000). Last judgment: the visionary biology of J. B. S. Haldane. *Journal of the History of Biology*, 33(3), pp. 457–491.

Adams, Mark B. (2009). Eugenics. In: Ravitsky, V., Fiester, A. and Caplan, A, eds. *The Penn Center Guide to Bioethics*. New York: Springer Pub, pp. 371–382.

Adams, Mark B., Allen, G. and Weiss, S. (2005). Human heredity and politics: a comparative institutional study of the Eugenics Record Office at Cold Spring Harbor (United States), the Kaiser Wilhelm Institute for Anthropology, Human Heredity, and Eugenics (Germany), and the Maxim Gorky Medical Genetics Institute (USSR). *Osiris*, 20(1), pp. 232–262.

Alaimo S. and Hekman S. eds. (2008). *Material Feminisms*. Bloomington/Indianapolis: Indiana University Press.

Alcoff, L. and Potter, E. (1993). *Feminist Epistemologies*. New York: Routledge.

Allen, G. (1975). *Life Science in the Twentieth Century*. New York: Cambridge University Press.

Allen, G. (1976). Genetics eugenics and society: externalist in contemporary history of science. *Social Studies of Science*, 6(1), pp. 105–122.

Allen, G. (1978). *Thomas Hunt Morgan: The Man and His Science*. Princeton, NJ: Princeton University Press.

Allen, G. (1979a) The rise and spread of the classical school of heredity 1910–1930; development and influence of the Mendelian chromosome theory. In: N. Reingold, ed. *Science in the American Context: New Perspectives*. Washington DC: Smithsonian, pp. 209–228.

Allen, G. (1979b). Naturalists and experimentalists: the genotype and the phenotype. *Stud Hist Biol*, 3, pp. 179–209.

Allen, G. (1987). The role of experts in scientific controversy. In: Tristram Engelhardt, H. and Arthur L. Caplan, eds., *Scientific Controversies: Case Studies in the Resolution and Closure of Disputes in Science and Technology*. Cambridge: Cambridge University Press, pp. 169–202.

Allen, G. (2011), Eugenics and modern biology: critiques of eugenics, 1910–1945. *Annals of Human Genetics*, 75, pp. 314–325. doi: 10.1111/j.1469-1809.2011.00649.x.

Almond, D. (2006). Is the 1918 influenza pandemic over? Long-term effects of in utero influenza exposure in the post-1940 U.S. population. *Journal of Political Economy*, 114, pp. 672–712, http://ideas.repec.org/a/ucp/jpolec/v114y2006i4p672-712.html.

Almond, D. and Currie, J. (2011). Killing me softly: the fetal origins hypothesis. *Journal of Economic Perspectives*, 25, pp. 153–172, http://www.aeaweb.org/articles.php? doi: 10.1257/jep. 25.3.153.

Amundson, R. (1998). Typology reconsidered: two doctrines on the history of evolutionary biology. *Biology and Philosophy*, 13, pp. 153–177.

Amundson, R. (2005). *The Changing Role of the Embryo in Evolutionary Thought*. Cambridge: Cambridge University Press.

Anderson, C. (2003). *Eyes Off the Prize: The United Nations and the African American Struggle for Human Rights, 1944–1955*. Cambridge, UK: Cambridge University Press.

Ankeny, R. and Leonelli, S. (2015). Valuing data in postgenomic biology. In: Richardson, S. and Stevens, H., eds. *Postgenomics: Perspectives on Biology after the Genome*. Durham, NC: Duke University Press, pp. 126–149.

Annas, G. (2005). *American Bioethics: Crossing Human Rights and Health Law Boundaries*. Oxford: Oxford University Press.

Annas, G. (2010). *Worst Case Bioethics: Death, Disaster, and Public Health*. New York: Oxford University Press.

Annas, G. and Grodin, M. (1992). *The Nazi Doctors and the Nuremberg Code: Human Rights in Human Experimentation*. New York: Oxford University Press.

Anway, M., Cupp, A., Uzumcu, M., and Skinner, M. (2005). Epigenetic transgenerational actions of endocrine disruptors and male fertility. *Science*, 308, pp. 1466–1469.

Armstrong, L. (2014). *Epigenetics*. London: Garland Science.

Arnhart, L. (1998). *Darwinian Natural Right. the Biological Ethics of Human Nature*, Albany NY: State University of New York Press.

Arnhart, L. (2005). *Darwinian Conservatism*, Exeter: Imprint Academic.

Ayala, F. J. (1976). Theodosius Dobzhansky: the man and the scientist. *Annual Review of Genetics*, 10, pp. 1–6.

Ayala, F. J. (1985). Theodosius Dobzhansky 1900–1975. *Biogr. Mem. Natl. Acad. Sci.*, 55, pp. 163–213.

Babkov, V. (2013). *The Dawn of Human Genetics*. Fet, V., trans., Schwartz, J., ed. Cold Spring Harbor, NY: Cold Spring Harbor Laboratory Press.

Badyaev, A. and Uller, T. (2009). Parental effects in ecology and evolution: mechanisms, processes, and implications. *Philso. Trans. R. Soc. Lond. B Biol. Sci.*, 364, pp. 1169–1177.

Baiter, M. (2000). Genetics: was Lamarck just a little bit right?. *Science*, 7(April), p. 38.

Bajema, C. J. (1976). *Eugenics: Then and Now*. New York: Halsted Press.

Bannister, R. (1988). *Social Darwinism: Science and Myth in Anglo-American Thought*. 2nd ed. Philadelphia, PA: Temple University Press.

Barad, K. (2003). Posthumanist performativity: toward an understanding of how matter comes to matter. *Signs*, 28 (3), pp. 801–831.

Barber, B. (1952). *Science and the Social Order*. Glencoe, IL: Free Press.

Barkan, E. (1992). *The Retreat of Scientific Racism: Changing Concepts of Race in Britain and the United States between the World Wars*. Cambridge: Cambridge University Press.

Barker, D. (1989). The biology of stupidity: genetics, eugenics and mental deficiency in the inter-war years. *The British Journal for the History of Science*, 22(03), p. 347.

Barker, D. J. P. (1995). Fetal origins of coronary heart disease. *British Medical Journal*, 311, pp. 171–174.

Barker, D. J. P. (1998). *Mothers, Babies, and Health in Later Life*. Edinburgh: Churchill Livingstone.

Barker, D. J. P and C. Osmond (1986). Infant mortality, childhood nutrition, and ischaemic heart disease in England and Wales. *The Lancet*, 327(8489), pp. 1077–1081.

Barkow, J. H., Cosmides L., and Tooby, J., eds. (1992). *The Adapted Mind: Evolutionary Psychology and the Generation of Culture*. Oxford & New York: Oxford University Press.

Barnes, B. (1974). *Scientific Knowledge and Sociological Theory*. London: Routledge & Kegan Paul.

Barnes, B. and Dupré, J. (2008). *Genomes and What to Make of Them*. Chicago: University of Chicago Press.

Barnes, B. and Shapin S., ed. (1979). *Natural Order: Historical Studies of Scientific Culture*. Beverly Hills, Calif.: Sage Publications.

Barnes, B., Bloor, D. and Henry, J. (1996). *Scientific Knowledge: A Sociological Analysis*, Chicago: Chicago UP.

Barrett, D. and Kurzman, C. (2004). Globalizing social movement theory: the case of eugenics. *Theory and Society*, 33(5), pp. 487–527.

Bashford A. (2010). *Epilogue*: where did *eugenics* go? In: Bashford A. and Levine P., eds. *The Oxford Handbook of the History of Eugenics*. New York: Oxford University Press.

Bashford A. and Levine P., eds. (2010). *The Oxford Handbook of the History of Eugenics*. New York: Oxford University Press.

Bateson, W. (1913). *Mendel's Principles of Heredity*. Cambridge: Cambridge University Press.

Bauman, Z. (1989). *Modernity and the Holocaust*. Ithaca, NY: Cornell University Press.

Beatty, J. (1987). Weighing the risks: stalemate in the classical/balance controversy. *Journal of the History of Biology*, 20(3), pp. 289–319.

Beatty, J. (1992). Julian Huxley and the evolutionary synthesis. In: Kenneth Waters, C. and Van Helden, Albert, eds. *Julian Huxley: Biologist and Statesman of Science*. Houston: Rice University Press.

Beatty, J. (1994). Dobzhansky and the biology of democracy: the moral and political significance of genetic variation. In: Adams, M., ed. *The Evolution of Theodosius Dobzhansky: Essays on His Life and Thought in Russia and America*. Princeton, NJ: Princeton University Press.

Bechtel, W. (2008). *Mental Mechanisms: Philosophical Perspectives on Cognitive Neuroscience*. New York: Routledge.

Bennett J. (2010). *Vibrant Matter: A Political Ecology of Things*. Durham/London: Duke University Press.
Beurton, P., Falk, R., and Rheinberger, H., eds. (2003). *The Concept of the Gene in Development and Evolution: Historical and Epistemological Perspectives*. Cambridge: Cambridge University Press.
Biémont, C. and Vieira, C. (2006). Genetics: junk DNA as an evolutionary force. *Nature*, 443(7111), pp. 521–524.
Bird, A. and Macleod, D. (2004). Reading the DNA methylation signal. *Cold spring Harb. Symp. Quant. Biol.*, 69, pp. 113–118.
Bird, A. (2007). Perceptions of epigenetics. *Nature*, 447, pp. 396–398.
Blacher, L. (1982). *The Problem of Inheritance of Acquired Characters: A History of a Priori and Empirical Methods Used to Find a Solution*. Churchill, F., trans. New Dehli: Amerind.
Black, E. (2003). *War against the Weak: Eugenics and America's Campaign to Create a Master Race*. New York: Four Walls Eight Windows.
Blacker, C. P. (1952a). "Eugenics" Experiment Conducted by the Nazis on Human Subjects. *The Eugenics Review*, 04/1952; 44(1), pp. 9–19.
Blacker, C. (1952b). *Eugenics: Galton and after.* Cambridge, Mass.: Harvard University Press.
Blackmon, D. A. (2009). *Slavery by Another Name: The Re-Enslavement of Black Americans from the Civil War to World War II.*. Anchor Books: New York.
Blane D., Kelly-Irving M., Errico A., Bartley M., and Montgomery S. (2013). Social-biological transitions: how does the social become biological? *Longitud Life Course Stud Int J*, 4(2), pp. 136–146.
Bliss C., (2015). Defining health justice in the postgenomic era. In: Richardson, S. and Stevens, H., eds. *Postgenomics: Perspectives on Biology after the Genome*. Durham, NC: Duke University Press, pp. 174–190.
Bloomfield, P. (1949). The eugenics of the utopians: the Utopia of the eugenists. *Eugenics Review*, 40, pp. 191–199.
Bloor, D. (1976). *Knowledge and Social Imagery*. London: Routledge and Kegan & Paul.
Blue, G. (2001). Scientific humanism and the founding of UNESCO. *Comparative Criticism*, 23, pp. 173–200.
Boas, F. (1910). *Changes in Bodily Form of Descendants of Immigrants*. Washington, D.C.: Government. Printing Office.
Boas, F. and Benedict, R. (1938). *General Anthropology*. Boston: D.C. Heath and Co.
Bock, G. (1983). Racism and sexism in Nazi Germany: motherhood, compulsory sterilization, and the state. *Signs Women and Violence*, 8(3), pp. 400–421.
Bonah, C., Danion-Grilliat, A., Olff-Nathan, J., Schappacher, N. (2006). *Nazisme, science* et *médecine*. Paris: Glyphe.
Bonduriansky, R. and Day, T. (2009). Nongenetic inheritance and its evolutionary implications. *Annu. Rev. Ecol. Evol. Syst.*, 40, pp. 103–125.
Bonduriansky, R. (2012). Rethinking heredity, again. *Trends in Ecology & Evolution*, 27(6), pp. 330–336.
Borrello, M. (2010). *Evolutionary Restraints: The Contentious History of Group Selection*. Chicago: University of Chicago Press.
Bowker, G. and Star, S. (1999). *Sorting Things Out: Classification and Its Consequences*. Cambridge, Mass.: MIT Press.

Bowler, P. (1983). *The Eclipse of Darwinism: Anti-Darwinian Evolution Theories in the Decades around 1900*. Baltimore: Johns Hopkins.

Bowler, P. (1984). E. W. MacBride's Lamarckian eugenics and its implications for the social construction of scientific knowledge. *Annals of Science*, 41(3), pp. 245–260.

Bowler, P. (1986). *Theories of Human Evolution: A Century of Debate, 1844–1944*. Baltimore: Johns Hopkins University Press.

Bowler, P. (1988). *The Non-Darwinian Revolution: Reinterpreting a Historical Myth*. Baltimore: Johns Hopkins University Press.

Bowler, P. (1989). *The Mendelian Revolution: The Emergence of Hereditarian Concepts in Modern Science and Society*. Baltimore: Johns Hopkins University Press.

Bowler, P. (1995). Social metaphors in evolutionary biology: 1870–1930. The wider dimension of social Darwinism. In: Maasen, S., Mendelsohn, E. and Weingart, P., eds. *Biology as Society, Society as Biology: Metaphors*, Netherlands: Springer, pp. 107–126.

Bowler, P. (2009). *Evolution, the History of an Idea*. Berkeley: University of California Press [third edition].

Boyd, W. (1950). *Genetics and the Races of Man. An Introduction to Modern Physical Anthropology*. Boston: Little, Brown and Co.

Boyd, W. (1953). The contributions of genetics to anthropology. In: Kroeber, A., ed. *Anthropology Today; An Encyclopedic Inventory*. Chicago: University of Chicago Press, pp. 448–506.

Brannigan, A. (1979). The reification of Gregor Mendel. *Social Studies of Science*, 9, pp. 423–454.

Brannigan, A. (1981). *The Social Basis of Scientific Discovery*. Cambridge: Cambridge University Press.

Brattain, M. (2007). Race, racism, and antiracism: *UNESCO* and the politics of presenting science to the postwar public. *The American Historical Review*, 112(5), pp. 1386–1413.

Breslau, D. (2007). The American Spencerians: theorizing a new science. In: Calhoun, C. ed., *Sociology in America: A History*. Chicago: University of Chicago Press, pp. 39–62.

Brewer, H. (1935). Eutelegenesis. *Eugenics Review*, 27, p. 121.

Brown, D. E. (1991). *Human Universals*. New York City: McGraw-Hill.

Brookes, M. (2004). *Extreme Measures: The Dark Visions and Bright Ideas of Francis Galton*. New York: Bloomsbury USA.

Brooks, W. (1883). *The Law of Heredity: A Study of the Cause of Variation, and the Origins of Living Organisms*. Baltimore, MD: J. Murphy.

Brown, D. E. (1991). *Human Universals*, New York: McGraw-Hill.

Bruinius, H. (2007). *Better for All the World: The Secret History of Forced Sterilization and America's Quest for Racial Purity*. New York: Vintage Books.

Buchanan, A., Brock, D., Daniels, N., and Wikler, D. (2000). *From Chance to Choice: Genetics and Justice*. Cambridge: Cambridge University Press.

Bulmer, M. (2004a). *Francis Galton: Pioneer of Heredity and Biometry*. Baltimore, MD: Johns Hopkins University Press.

Bulmer, M. (2004b). Did Jenkin's Swamping Argument Invalidate Darwin's Theory of Natural Selection, *The British Journal for the History of Science*, 37(3), pp. 281–297.

Buller, D. (2006). *Adapting Minds: Evolutionary Psychology and the Persistent Quest for Human Nature*. Cambridge, MA: MIT Press.

Burleigh, M. and Wippermann, W. (1993). *The Racial State: Germany, 1933–1945.* Cambridge: Cambridge University Press.
Burkhardt, Jr., R. W. (1977). *The Spirit of System: Lamarck and Evolutionary Biology.* Cambridge, MA/London: Harvard University Press.
Burkhardt, Jr., R. W. (1980). Lamarckism in Britain and the United States. In: Mayr, E. and Provine, W. B., eds. *The Evolutionary Synthesis*, pp. 343–352, Cambridge, Mass: Harvard University Press.
Burkhardt, Jr., R. W. (2013). Lamarck, Evolution, and the Inheritance of Acquired Characters. *Genetics*, 194(4), pp. 793–805.
Burke, R. (2010). *Decolonization and the Evolution of International Human Rights.* Philadelphia, PA: University of Pennsylvania Press.
Burrow, J. (1966). *Evolution and Society: A Study in Victorian Social Theory.* Cambridge: Cambridge University Press.
Bush, V. (1945). *Science – The Endless Frontier: A Report to the President on a Program for Postwar Scientific Research.* Reprinted 1990. Washington, DC: National Science Foundation [earlier reprints].
Butler, S. (1880). *Unconscious Memory.* London: Bogue.
Bygren, L., Kaati, G., and Edvinsson, S. (2001). Longevity determined by ancestors' overnutrition during their slow growth period. *Acta Biotheoretica*, 49, pp. 53–59.
Bynum, W. (1984). Alcoholism and degeneration in 19th century European medicine and psychiatry. *British Journal of Addiction*, 79(1), pp. 59–70.
Calhoun, C., ed. (2007). *Sociology in America: A History.* Chicago: University of Chicago Press.
Callender, L. A. (1988). Gregor Mendel: an opponent of descent with modification. *History of Science*, 26, pp. 41–75.
Calvocoressi, P. (1991). *World Politics Since 1945.* London: Longman.
Campbell, T. and Sitze, A., ed. (2013). *Biopolitics: A Reader.* Durham NC: Duke University Press.
Caplan, A. (1992). How did medicine go so wrong? In: Caplan, A., ed. *When Medicine Went Mad.* Totowa, N. J.: Humana Press.
Carlson, E. (1981). *Genes, Radiation, and Society: The Life and Work of H. J. Muller.* Ithaca: Cornell University Press.
Carlson, E. (1987). Eugenics and basic genetics in H. J. Muller's approach to human genetics. *History and Philosophy of the Life Sciences*, 9(1), pp. 57–78.
Carlson, E. (2001). *The Unfit. A History of a Bad Idea.* Cold Spring Harbor, NY: Cold Spring Harbor Press.
Carlson, E. A. (2009). *Hermann Joseph Muller 1890–1967: A Biographical Memoir.* Washington DC: National Academy of Science.
Carlson, E. A. (2011). Speaking out about the social implications of science: the uneven legacy of H. J. Muller, *Genetics* (January), 187(1), pp. 1–7.
Carr-Saunders, A. (1911). Some recent eugenic work. *Economic Review* (January), pp. 19–27.
Carr-Saunders, A. (1913). A criticism of eugenics, *The Eugenics review*, 10, pp. 214–33.
Cassin, R. (1968). Noble Prize Lecture. Available at: www.nobelprize.org/nobel_prizes/peace/laureates/1968/cassin lecture.html.
Ceccarelli, L. (2003). *Shaping Science with Rhetoric: The Cases of Dobzhansky, Schrodinger, and Wilson*, Chicago: University of Chicago Press.

Champagne, F. (2008). Epigenetic mechanisms and the transgenerational effects of maternal care. *Frontiers in Neuroendocrinology*, 29, pp. 386–397.

Champagne, F. (2010). Epigenetic perspectives on development: evolving insights on the origins of variation, *Developmental Psychobiology*, 52, pp. e1–e3.

Champagne, F. and Curley, J. (2008). Maternal regulation of estrogen receptor a methylation. *Current Opinion in Pharmacology*, 8, pp. 735–739.

Champagne, F., and Curley, J. (2009). Epigenetic mechanisms mediating the long-term effects of maternal care on development. *Neuroscience and Biobehavioral Reviews*, 33, pp. 593–600.

Charney, E. (2012). Behavior genetics and postgenomics. *Behavioral and Brain Sciences*, 35(5), pp. 331–358.

Chesterton, G. K. (1922). *Eugenics and Other Evils*, London: Cassell and Company, Limited.

Chong, S., and Whitelaw, E. (2004). Epigenetic germline inheritance. *Current Opinion in Genetics & Development*, 14, pp. 692–696.

Chung, C. (2003). On the origin of the typological/population distinction in Ernst Mayr's changing views of species, 1942–1959. *Stud. Hist. Phil. Biol. & Biomed. Sci.*, 34, pp. 277–296.

Churchill, F. (1968). August Weismann and a break from tradition. *Journal of the History of Biology*, 1(1), pp. 91–112.

Churchill, F. (1974). William Johannsen and the Genotype Concept. *Journal of the History of Biology*, 7, pp. 5–30.

Churchill, F. (1976). Rudolf Virchow and the pathologist's criteria for the inheritance of acquired characteristics. *Journal of the History of Medicine and Allied Sciences*, 31(2), pp. 117–148.

Churchill, F. (1987). From hereditary theory to Vererbung: the transmission problem, 1850–1915. *Isis*, 78, pp. 337–364.

Churchill, F. (2015). *August Weismann: Development, Heredity, and Evolution*. Cambridge, MA: Harvard UP.

Cock, A., Forsdyke D. R. (2008). *Treasure Your Exceptions: The Science and Life of William Bateson*. New York: Springer.

Coffin, J. (2003). Heredity, milieu and sin: the works of Bénédict Augustin Morel (1809–1873). In: *A Cultural History of Heredity*. Berlin: Max-Planck Institut für Wissenschaftsgeschichte, pp. 153–164.

Collins, H. (1974). The *TEA* Set: Tacit Knowledge and Scientific Networks. *Science Studies*, 4, pp. 165–186.

Collins, H. (2010). *Tacit and Explicit Knowledge*. Chicago: Chicago University Press.

Collins, J., Rankin K., and David R. (2011). African American women's lifetime upward economic mobility and preterm birth: the effect of fetal programming. *Am J Public Health*, 101(4), pp. 714–719.

Collopy J. S. (2015). Race relationships: collegiality and demarcation in physical anthropology. *Journal of the History of the Behavioural Sciences* (April), pp. 1–24.

Cook, G. M. (1999). Neo-Lamarckian experimentalism in America: origins and consequences. *The Quarterly Review of Biology*, 74(4), pp. 417–437.

Cooke, K. J. (1998). The limits of heredity: nature and nurture in American eugenics before 1915. *Journal of the History of Biology*, 30(2), p. 263.

Coole, D. and Frost, S. (2010). *New Materialisms: Ontology, Agency and Politics*. Durham NC: Duke UP.

Cooper, J. (2008). *Raphael Lemkin and the Struggle for the Genocide Convention.* Basingstoke, England: Palgrave Macmillan.
Corsi, P. (1988). *The Age of Lamarck: Evolutionary Theories in France, 1790–1830.* Berkeley, CA: University of California Press.
Corsi, P. (2011). Jean-Baptiste Lamarck. In: Gissis, S. and Jablonka, E., eds. *Transformations of Lamarckism: From Subtle Fluids to Molecular Biology*. Cambridge, Mass.: MIT Press.
Cowan, R. (1972a). Francis Galton's contribution to genetics. *Journal of the History of Biology*, 5(2), pp. 389–412.
Cowan, R. (1972b). Francis Galton's statistical ideas: the influence of eugenics. *Isis*, 63(4), p. 509.
Cowan, R. (1977). Nature and nurture: the interplay of biology and politics in the work of Francis Galton. *Stud Hist Biol.*, 1, pp. 133–208.
Cowan, R. (1985). *Sir Francis Galton and the Study of Heredity in the Nineteenth Century.* New York: Garland Pub.
Cowan, R. (2008). *Heredity and Hope: The Case for Genetic Screening.* Cambridge, Mass: Harvard UP.
Crackanthorpe, M. (1909). The eugenic field. *Eugen Rev.*, 1(1), pp. 11–25.
Cravens, H. (1971). The abandonment of evolutionary social theory in America: the impact of academic professionalization upon American sociological theory, 1890–1920. *American Studies*, 12, pp. 5–20.
Cravens, H. (1978). *The Triumph of Evolution: American Scientists and the Heredity-Environment Controversy*. Philadelphia: University of Pennsylvania Press.
Crean, A. J., Kopps, A., and Bonduriansky, R. (2014). Revisiting telegony: offspring inherit an acquired characteristic of their mother's previous mate. *Ecology Letters*, 17, pp. 1545–1552.
Crick, F. H. C. (1958). On protein synthesis. *Symposia of the Society for Experimental Biology*, 12, pp. 138–163.
Crick, F. H. C. (1970). Central dogma of molecular biology. *Nature*, 227(5258), pp. 561–563.
Crook D. P. (1994). *Darwinism, War and History.* Cambridge UK: Cambridge University Press.
Crook, D. P. (2007). *Darwin's Coat-tails: Essays on Social Darwinism.* New York; Peter Lang.
Curley, J. P., Mashoodh, R., and Champagne, F. A. (2011). Epigenetics and the origins of paternal effects. *Hormones and Behavior*, 59, pp. 306–314.
Curtin, P. D. (1992). The slavery hypothesis for hypertension among African Americans: the historical evidence. *American Journal of Public Health*, 82(12), pp. 1681–1686.
Danchin, E. (2013). Avatars of information. *Trends in Ecology and Evolution*, 28, pp. 351–358.
Danchin, E., Charmantier, A., Champagne, F., Mesoudi, A., Pujol B., and Blanchet, S. (2011). Beyond DNA: integrating inclusive inheritance in an extended theory of evolution. *Nature Reviews Genetics*, 12, pp. 475–486.
Darlington, C. (1959). *Darwin's Place in History*. London: Blackwell.
Darwin, C. (1859). *On the Origin of Species by Means of Natural Selection, or the Preservation of Favoured Races in the Struggle for Life.* London: John Murray.
Darwin, C. (1875). *Variation in Animals and Plants under Domestication.* 2nd edition. London: J. Murray.

Darwin, F., ed. (1909). *The Foundations of the Origin of Species, Two Essays Written in 1842 and 1844*. Cambridge: Cambridge University Press.
Darwin, L. (1912). First Step Toward Eugenic Reform. *The Eugenics Review*, 4(1), pp. 26–38.
Daston, L. (1994). Historical epistemology. In: Chandler, J., A. I. Davidson, and Harootunian, H., eds., *Questions of Evidence*, Chicago: University of Chicago Press, pp. 282–289.
Daston, L. and Park, K. (1998). *Wonders and the Order of Nature*. New York: Zone Books.
Daston, L., ed. (2000). *Biographies of Scientific Objects*. Chicago: University of Chicago Press.
Daston, L. (2004). The Morality of Natural Orders: The Power of Media: In Peterson, G. B., ed. *The Tanner Lectures on Human Values*, vol. 24. Salt Lake City: University of Utah Press, pp. 371–392.
Daston, L. and Vidal, F. (2007). *The Moral Authority of Nature*. Chicago: Chicago University Press.
Daston, L. and Galison, P. (2010). *Objectivity*. New York: Zone Books.
Davenport, C. (1911). *Heredity in Relation to Eugenics*. New York: Holt and Co.
Dawkins, R. (2010). *"The Information Challenge" in A Devil's Chaplain: Selected Writings*. New York: First Mariner Books.
Daxinger, L., and Whitelaw, E. (2010). Transgenerational epigenetic inheritance: more questions than answers. *Genome Research*, 20, pp. 1623–1628.
Daxinger, L., and Whitelaw, E. (2012). Understanding transgenerational epigenetic inheritance via the gametes in mammals. *Nature Reviews Genetics*, 13, pp. 153–162.
Degler, C. (1989). *Culture versus Biology in the Thought of Franz Boas and Alfred L. Kroeber*, German Historical Institute, Annual Lecture Series, New York: Berg.
Degler, C. (1991). *In Search of Human Nature: The Decline and Revival of Darwinism in American Social Thought*. New York: Oxford University Press.
Deichmann, U. and Müller-Hill, B. (1994). Biological research at universities and Kaiser Wilhelm Institutes in Nazi Germany. In: Renneberg, M. and Walker, M., eds. *Science, Technology, and National Socialism*. Cambridge, England: Cambridge University Press, pp. 160–183.
Deichmann, U. (1996). *Biologists under Hitler*. Dunlap, D., trans. Cambridge, Mass.: Harvard University Press.
Delage, Y. and Goldsmith, M. (1909/1912). *The Theories of Evolution*. New York: Huebsch (original 1909).
Demerec, M. (1950). *Origin and Evolution of Man*. Cold Spring Harbor Symposia on Quantitative Biology, vol. 15. Cold Spring Harbor, NY: Biological Laboratory.
Depew, D. (2013). Challenging Darwinism: expanding, extending, or replacing the modern evolutionary synthesis. In: Ruse, M., ed. *The Cambridge Encyclopedia of Darwin*. Cambridge: Cambridge University Press.
Depew, D. and Weber, B. H. (2011). The fate of Darwinism: evolution after the modern synthesis. *Biological Theory*, 6(1), pp. 89–102.
Desmond, A. (1989). *The Politics of Evolution: Morphology, Medicine, and Reform in Radical London*. Chicago: University of Chicago Press.
Desmond, A. and Moore, J. (2009). *Darwin's Sacred Cause: How a Hatred of Slavery Shaped Darwin's Views on Human Evolution*. London: Allen Lane.

de Chadarevian, S. (2014). Chromosome surveys of human populations: between epidemiology and anthropology. *Studies in History and Philosophy of Science Part C: Studies in History and Philosophy of Biological and Biomedical Sciences*, 47, pp. 87–96.
de Jong-Lambert, W. and Krementsov, N. (2011). On labels and issues: the Lysenko Controversy and the Cold War. *Journal of the History of Biology*, pp. 373–388.
de Jong Lambert, W. (2012). *The Cold War Politics of Genetic Research. An Introduction to the Lysenko Affair*. Springer.
Dias, B. and Ressler, K. (2014). Parental olfactory experience influences behavior and neural structure in subsequent generations. *Nature Neuroscience*, 17(1), pp. 89–96.
Diez Roux, A., Jacobs, D., and Kiefe, C. (2002). Neighborhood characteristics and components of the insulin resistance syndrome in young adults: the coronary artery risk development in young adults (CARDIA) study. *Diabetes Care*, 25, pp. 1976–1982.
Dikötter, F. (1998). Race culture: recent perspectives on the history of eugenics. *American Historical Review*, 103, pp. 467–478.
Dobzhansky, T. (1937). *Genetics and the Origin of Species*. New York: Columbia University Press.
Dobzhansky, T. (1945). An outline of politico-genetics. *Science*, 102(2644), pp. 234–236.
Dobzhansky, T. and Montagu, M. (1947). Natural selection and the mental capacities of mankind. *Science*, 105(2736), pp. 587–590.
Dobzhansky, T. and Wallace B. (1954). The problem of adaptive differences in human populations. *American Journal of Human Genetics*, 6(2).
Dobzhansky, T. (1955). A review of some fundamental concepts and problems of population genetics, *Cold Spr. Harbor Symp. Quant. Biol.*, 20, pp. 1–15.
Dobzhansky, T. (1956). *The Biological Basis of Human Freedom*. N. Y.: Columbia Univ. Press.
Dobzhansky, T. (1961). The bogus "science" of race prejudice [Review of *Race and reason: A Yankee view* by C. Putnam]. *Journal of Heredity*, 52, pp. 189–190.
Dobzhansky, T. (1962). *Mankind Evolving*. New Haven: Yale University Press.
Dobzhansky, T. (1963). A debatable account of the origin of races [Review of *The origin of races* by C. S. Coon]. *Scientific American*, 208, pp. 169–172.
Dobzhansky, T. (1964). Evolution: organic and superorganic. *Bulletin of The Atomic Scientists*, 20(5), pp. 4–8.
Dobzhansky, T. (1967). *The Biology of Ultimate Concern*. New York: New American Library.
Dolinoy, D., and Jirtle, R. (2008). Environmental epigenomics in human health and disease. *Environmental and Molecular Mutagenesis*, 49(1), pp. 4–8.
Douglas, R. (2002). Anglo-Saxons and Attacotti: the racialization of Irishness in Britain between the World Wars. *Ethnic and Racial Studies*, 25(1), pp. 40–63.
Dowbiggin, I. (1991). *Inheriting Madness: Professionalization and Psychiatric Knowledge in 19th Century France*. Berkeley: University of California Press.
Drake, A. and Liu, L. (2010). Intergenerational transmission of programmed effects: public health consequences. *Trends in Endocrinology and Metabolism*, 21(4), pp. 206–213.
Dronamraju, K. (1995). *Haldane's Daedalus Revisited*. New York: Oxford University Press.

Dudziak, M. (2000). *Cold War Civil Rights: Race and the Image of American Democracy*. Princeton, NJ: Princeton University Press.

Duello, M. T. (2010). Misconceptions of "race" as a biological category: then and now. In: Sheldon Rubenfeld, ed., *Medicine After the Holocaust: From the Master Race to the Human Genome and Beyond*. New York: Palgrave MacMillan, pp. 37–48.

Dulac, C. (2010). Brain function and chromatin plasticity. *Nature*, 465, pp. 728–735.

Dunn, L. and Dobzhansky T. (1952). *Heredity, Race, and Society*. Enlarged edition. New York: New American Library.

Dupras, C., Ravitsky, V., and Williams-Jones, B. (2014). Epigenetics and the environment in bioethics. *Bioethics*, 28(7), pp. 327–334.

Dupré, J. (2012). *Processes of Life: Essays in the Philosophy of Biology*. Oxford: Oxford University Press.

Duster, T. (2003). *Backdoor to eugenics*. New York: Routledge.

(The) Economist. (2013). Poisoned Inheritance, December 14. http://www.economist.com/news/science-and-technology/21591547-lack-folate-diet-male-mice-reprograms-their-sperm-ways.

Eichler E. E. et al. (2010). Missing heritability and strategies for finding the underlying causes of complex disease. *Nat Rev Genet*, 11(6), pp. 446–450.

Ekberg, M. (2007). The old eugenics and the new genetics compared. *Social History of Medicine*, 20(3), pp. 581–593.

Elderton, E. and Pearson, K. (1910). *A First Study of the Influence of Parental Alcoholism on the Physique and Ability of the Offspring*. London: Dulau and Co. Ltd.

Ellis, H. (1911). *The Problem of Race-Regeneration*. New York: Moffat, Yard, and Company.

Emanuel, I. (1986). Maternal health during childhood and later reproductive performance. *Ann N Y Acad Sci.*, 477, pp. 27–39.

ENCODE Project Consortium (2007). Identification and analysis of functional elements in 1% of the human genome by the ENCODE pilot project. *Nature*, 447, pp. 799–816.

ENCODE Project Consortium (2012). An integrated encyclopedia of DNA elements in the human genome. *Nature*, 489, pp. 57–74.

Esposito, M. (2011). Utopianism in the British evolutionary synthesis. *Studies in History and Philosophy of Science Part C*, 42(1), pp. 40–49.

Esposito, M. (2013). Weismann versus Morgan revisited: clashing interpretations on animal regeneration. *Journal of the History of Biology*, 46, pp. 511–541.

Esposito, R. (2008a). *Bíos: Biopolitics and Philosophy*. Minneapolis: University of Minnesota Press.

Esposito, R. (2008b). Totalitarianism or biopolitics? Concerning a philosophical interpretation of the twentieth century. *Critical Inquiry*, 34, pp. 633–644.

Falk, R. (1995). The struggle of genetics for independence. *Journal of the History of Biology*, 28(2), pp. 219–246.

Falk, R. (2003). The gene: a concept in tension. In: Beurton, P., Falk, R., and Rheinberger H., eds. *The Concept of the Gene in Development and Evolution: Historical and Epistemological Perspectives*, Cambridge: Cambridge University Press, pp. 317–348.

Fancher, R. E. (2009). Scientific cousins: the relationship between Charles Darwin and Francis Galton. *Am Psychol* (Feb–Mar), 64(2), pp. 84–92.

Farber, P. (2010). *Mixing Races: From Scientific Racism to Modern Evolutionary Ideas.* Baltimore: Johns Hopkins University Press.

Farrall, L. (1970). The origins and growth of the English eugenics movement, 1865–1925. PhD dissertation, Indiana University.

Fassin, D. (2007). Humanitarianism as a politics of life. *Public Culture,* 19(3), pp. 499–520.

Fassin, D. (2009a). Another politics of life is possible. *Theory Culture and Society,* 26(5), pp. 44–60.

Fassin, D. (2009b). Les économies morales revisitées. *Annales. Histoire, Sciences Sociales,* 64(6), pp. 1237–1266.

Fassin, D. (2015). Moral Economy Redux: A Critical Reappraisal. Text presented at the Institute for Advanced Study, SSS WIT seminar, Princeton New Jersey, April.

Fausto-Sterling, A. (1985). *Myths of Gender: Biological Theories about Women and Men,* New York: Basic Books.

Fausto-Sterling, A. (2000). *Sexing the Body: Gender Politics and the Construction of Sexuality.* New York: Basic Books.

Feil, R., and Fraga, M. (2012). Epigenetics and the environment: emerging patterns and implications. *Nature Review Genetics,* 13(2), pp. 97–109.

Fisher, R. (1922). Contribution to a discussion on the inheritance of mental qualities, good and bad. *Eugenics Review,* 14, pp. 210–213.

Fisher, R. (1922). The evolution of the conscience in civilized communities. *Eugenics Review,* 14, pp. 190–193.

Fisher, R. (1926). Eugenics: can it solve the problem of decay of civilizations? *Eugenics Review,* 18, pp. 128–136.

Fisher, R. (1930). *The Genetical Theory of Natural Selection.* Oxford: The Clarendon Press.

Fisher, R. (1931). The biological effects of family allowances. *Family Endowment Chronicle,* 1, pp. 21–25.

Fisher Box, J. (1978). *R. a. Fisher: The Life of a Scientist.* New York: Wiley.

Fleck, L. (1979). *The Genesis and Development of a Scientific Fact.* Trenn, T. J. and Merton, R. K., eds., foreword by Thomas Kuhn. Chicago: University of Chicago Press, (original 1935).

Forrest, D. (1974). *Francis Galton: The Life and Work of a Victorian Genius.* New York: Taplinger Pub. Co.

Forsdahl. A. (1977). Are poor living conditions in childhood and adolescence an important risk factor for arteriosclerotic disease? *British Journal of Preventive and Social Medicine,* 31(2), pp. 91–95.

Forsdahl. A. (1978). Living conditions in childhood and subsequent development of risk factors for arteriosclerotic heart disease. The cardiovascular survey in Denmark 1974–75. *Journal of Epidemiology and Community Health,* 32(1), pp. 34–37.

Fortun, M. (2015). Affective assemblages in genomics and postgenomics. In: Richardson, S. and Stevens, H., eds. *Postgenomics: Perspectives on Biology after the Genome.* Durham, NC: Duke University Press, pp. 32–55.

Foucault, M. (1970), *The Order of Things: An Archaeology of the Human Sciences.* New York: Pantheon Books (original 1966).

Foucault, M. (1972). *The Archaeology of Knowledge.* New York: Pantheon Books.

Foucault, M. (1978). *History of Sexuality,* vol. 1. An introduction. New York: Random House.

Foucault, M. (1984). Polemics, politics and problematization: an interview. In: Rabinow, P., ed. *The Foucault Reader*. Harmondsworth, Mx: Penguin.

Foucault, M. (1985). *The Use of Pleasure,* trans. R. Hurley. New York: Penguin.

Foucault, M. (1986). Kant on Enlightenment and revolution. *Economy and Society,* 15, pp. 88–96.

Foucault, M. (2002). *Society must be defended: Lectures at the Collège de France, 1975–76*. New York: Picador.

Foucault, M. (2008). *The Birth of Biopolitics*. New York: Palgrave Macmillan.

Fracchia, J. and Lewontin, R. (1999). Does Culture Evolve? *History and Theory,* 38(4), pp. 52–78.

Francis, R. (2011). *Epigenetics: The Ultimate Mystery of Inheritance*. New York: W.W. Norton.

Franklin, T., Russig, H., Weiss, I., Gräff, J., Linder, N., Michalon, A., et. al. (2010). Epigenetic transmission of the impact of early stress across generations. *Biol. Psychiatry,* 68, pp. 408–415.

Freeden, M. (1979). Eugenics and progressive thought: a study in ideological affinity. *The Historical Journal,* 22(3), pp. 645–671.

Frost, S. (2016). *Biocultural Creatures: Toward a New Theory of the Human*. Durham: Duke UP.

Fuentes, A. (2008). The new biological anthropology: bringing Washburn's new physical anthropology into 2010 and beyond. In: *Seventy-seventh annual meeting of the American Association of Physical Anthropologists*.

Fukuyama, F. (1999). *The Great Disruption: Human Nature and the Reconstitution of Social Order*. New York: Free Press.

Fukuyama, F. (2002). *Our Posthuman Future: Consequences of the Biotechnology Revolution*. New York: Strauss and Giroux.

Fukuyama, F. (2006). *After the Neocons. America at the Crossroads,* London: Profile.

Fukuyama, F. (2011). The *Origins of Political Order: From Prehuman Times to the French Revolution*. New York: Farrar Strauss and Giroux.

Fuller S. (2000). *Governance of Science: Ideology and the Future of the Open Society*. Open Court.

Fullwiley, D. (2008). The biologistical construction of race: "admixture" technology and the new genetic medicine. *Social Studies of Science,* 38(5), pp. 695–735.

Furedi, F. (1998). *The Silent War: Imperialism and the Changing Perception of Race*. New Brunswick, NJ: Rutgers University Press.

Galton, F. (1865). Hereditary talent and character. *McMillan's Magazine,* 12(1), pp. 157–166; 318–327.

Galton, F. (1869). *Hereditary Genius*. London: Macmillan & Co.

Galton, F. (1871). Experiments in Pangenesis, *Proceedings of the Royal Society of London,* 19, pp. 393–410.

Galton, F. (1874). *English Men of Science: Their Nature and Nuture*. London: Macmillan & Co.

Galton, F. (1876). A theory of heredity. *The Journal of the Anthropological Institute of Great Britain and Ireland,* 5, p. 329.

Galton, F. (1883). *Inquiries into Human Faculty and Its Development*. London: Macmillan & Co.

Galton, F. (1884). Measurement of character. *Psychometry*, 36, pp. 179–185.

Galton, F. (1904). Eugenics: its definition, scope and aims. *The American Journal of Sociology*, 10, pp. 1–25.

Galton, F. (1909). *Essays in Eugenics*. London: The Eugenics Education Society.

Gannett, L. (2001). Racism and human genome diversity research: the ethical limits of "population thinking." *Philosophy of Science*, 68, pp. 479–492.

Gayon, J. (1998). *Darwinism's Struggle for Survival: Heredity and the Hypothesis of Natural Selection*. Cambridge: Cambridge University Press.

Gayon, J. (2003). Do the biologists need the expression "human race": UNESCO 1950–51. In: J. Rozenberg, ed. *Bioethical and Ethical Issues Surrounding the Trials and Code of Nuremberg: Nuremberg Revisited*. Lewiston: Edwin Mellen Press, pp. 23–48.

Gazzaniga, M. S. (2005). *The Ethical Brain*, New York: The Dana Press.

Geronimus, A. (2013). Deep integration: letting the epigenome out of the bottle without losing sight of the structural origins of population health. *American Journal of Public Health*, 103, pp. S56–S63.

Gieryn, T. (1983). Boundary-work and the demarcation of science from non-science: strains and interests in professional ideologies of scientists. *American Sociological Review*, 48(6), p. 781.

Gieryn, T. (1999). *Cultural Boundaries of Science: Credibility on the Line*. Chicago: University of Chicago Press.

Gigerenzer, G., Swijtink, Z., Porter, T., Daston, L., Beatty, J., and Kruger, L. (1989). *Empire of Chance: How Probability Changed Science and Everyday Life*. Cambridge: Cambridge University Press.

Gilbert, N. and Mulkay, M. (1984). *Opening Pandora's Box: A Sociological Analysis of Scientists' Discourse*. Cambridge: Cambridge University Press.

Gilbert S. (1988). Cellular politics: Ernest Everett Just, Richard B. Goldschmidt, and the attempts to reconcile embryology and genetics. In: Rainger, R., Benson, K., and Maienschein, J. eds. *The American Development of Biology*. Philadelphia, PA: University of Pennsylvania Press, pp. 311–346.

Gilbert, S. (2003). The reactive genome. In: Muller, G. and Newman, S., eds. *Origination of Organismal Form: Beyond the Gene in Developmental and Evolutionary Biology*. Cambridge, Mass.: MIT Press, pp. 87–101.

Gilbert, S. (2011). In: Gissis, S. and Jablonka, E., eds. *Transformations of Lamarckism: From Subtle Fluids to Molecular Biology*. Cambridge, Mass.: MIT Press.

Gilbert, S. and Epel, D. (2009). *Ecological Developmental Biology: Integrating Epigenetics, Medicine, and Evolution*. Sunderland, Mass.: Sinaeur Associates, Inc.

Gilbert, S., Sapp J. and Tauber, A. (2012). A Symbiotic View of Life: We Have Never Been Individuals. *The Quarterly Review of Biology*, 87(4), pp. 325–341.

Gillette, A. (2007). *Eugenics and the Nature-Nurture Debate in the Twentieth Century*. New York: Palgrave Macmillan.

Gillham, N. (2001). *A Life of Sir Francis Galton: From African Exploration to the Birth of Eugenics*. Oxford: Oxford University Press.

Gingras, Y. (2010). Naming without necessity: on the geneaology and uses of the label "historical epistemology". *Revue de Synthese*, 131, pp. 439–454.

Gintis H. (2004). Towards the Unity of Human Behavioural Sciences. *Behavioural and Brain Sciences*, 3(1), pp. 37–57.

Gissis, S. (2002). Late nineteenth century Lamarckism and French sociology. *Perspectives on Science*, 10(1), pp. 69–122.
Gissis, S. and Jablonka, E. (2011). *Transformations of Lamarckism: From Subtle Fluids to Molecular Biology*. Cambridge, Mass.: MIT Press.
Glad, J. (2003). Hermann J. Muller's 1936 letter to Stalin. *Mankind Quarterly*, 43(3), pp. 305–319.
Glausiusz, J. (2014). Searching chromosomes for the legacy of trauma. *Nature Commentary*, 11 June accessed at http://www.nature.com/news/searching-chromosomes-for-the-legacy-of-trauma-1.15369.
Gliboff, S. (2005). "Protoplasm is Soft Wax in our Hands": Paul Kammerer and the Art of Biological Transformation. *Endeavour*, 29(4), pp. 162–167.
Gliboff, S. (2006). The case of Paul Kammerer: evolution and experimentation in the early twentieth century. *Journal of the History of Biology*, 39, pp. 525–563.
Gliboff, S. (2011). The golden age of Lamarckism, 1866–1926. In: Gissis, S. and Jablonka, E., eds. *Transformations of Lamarckism: from subtle fluids to molecular biology*. Cambridge, Mass.: MIT Press, pp. 45–55.
Gluckman, P. (2004). Living with the past: evolution, development, and patterns of disease. *Science*, 305(5691), pp. 1733–1736.
Gluckman, P. and Hanson, M. (2005). *The Fetal Matrix: Evolution, Development and Disease*. New York: Cambridge University Press.
Gluckman, P., Hanson, M., Bateson, P., Beedle, A., Law, C., Bhutta, Z., et. al. (2009). Towards a new developmental synthesis: adaptive developmental plasticity and human disease. *Lancet*, 33, pp. 1654–1657.
Gluckman, P., Hanson, M., and Buklijas, T. (2010). A conceptual framework for the developmental origins of health and disease. *J Dev Orig Health Dis.*, 1(1), pp. 6–18.
Gluckman, P., Hanson, M., Beedle, A., Buklijas, T., and Low, F. (2011). In: Hallgrimsson, B. and Hall, B., eds. *Epigenetics: Linking Genotype and Phenotype in Development and Evolution*. Berkeley: University of California Press, pp. 398–423.
Gluckman, P., Hanson, M. (2012). *Fat, Fate, and Disease. Why Exercise and Diet Are Not Enough*. Oxford UP: New York.
Gluckman, P., Low, F., and Hanson, M. (2013). Developmental epigenomics and metabolic disease. In: Jirtle, R. and Tyson, F., eds. (2013). *Environmental Epigenomics in Health and Disease*. Berlin: Springer, pp. 31–50.
Goddard, H. (1912). *The Kallikak Family: A Study in the Heredity of Feeblemindedness*. New York: Macmillan Company.
Goddard, H. (1914). *Feeble-Mindedness: Its Causes and Consequences*. New York: Macmillan.
Godfrey, K., Gluckman, P., and Hanson, M. (2010). Developmental origins of metabolic disease: life course and intergenerational perspectives. *Trends in Endocrinology & Metabolism*, 21(4), pp. 199–205.
Gökyiğit, E. (1994). The reception of Francis Galton's "Hereditary Genius" in the Victorian periodical press. *Journal of the History of Biology*, 27(2), pp. 215–240.
Gordin, M. (2009). *Red Cloud at Dawn: Truman, Stalin, and the End of the Atomic Monopoly*. New York: Farrar, Straus, and Giroux.
Gordin, M. (2012). How Lysenkoism became pseudoscience: Dobzhansky to Velikovsky. *Journal of the History of Biology*, 45(3), pp. 443–468.

Gordin, M. (2015). Lysenko Unemployed: Soviet Genetics after the Aftermath. Presentation at the Shelby Cullom Davis Center for Historical Studies, Princeton University, March 6.

Gormley, M. (2007). Geneticist L. C. Dunn: politics, activism and community. PhD dissertation, Oregon State University, Corvallis. http://search.proquest.com/docview/304819966.

Gormley, M. (2009a). The Roman campaign of '53 to '55: the Dunn family among a Jewish community. In: Farber, Paul and Cravens, Hamilton, eds. *Race and Science: Scientific Challenges to Racism in Modern America*. Corvallis: Oregon State University Press, pp. 95–129.

Gormley, M. (2009b). Scientific discrimination and the activist scientist: L. C. Dunn and the professionalization of genetics and human genetics in the United States. *Journal of the History of Biology*, 42, pp. 33–72.

Gorriaz, J., Gumpenberger, C., and Wieland, M. (2011). Galton 2011 revisited: a bibliometric journey in the footprints of a universal genius. *Scientometrics*, 88, pp. 627–652.

Gould, S. (1980). Shades of Lamarck, reprinted in *The Panda's Thumb: More Reflections in Natural History.*

Gould, S. (1981). *The Mismeasure of Man.* 2nd ed. (1996). New York: Norton.

Gould, S. (1991). The smoking gun of eugenics. In: Gould, S., ed. (1995). *Dinosaur in a Haystack: Reflections in Natural History.* New York: Harmony Books, pp. 296–308.

Gould, S. (1997). Evolution: the pleasures of pluralism. *New York Review of Books*, pp. 47–52.

Graham, L. (1972). *Science and Philosophy in the Soviet Union*. New York: Alfred Knopf.

Graham, L. (1977). Science and values: the eugenics movement in Germany and Russia in the 1920s. *The American Historical Review*, 82(5), p. 1133.

Graham, L. (1981). *Between Science and Values*. New York: Columbia University Press.

Graham, L. (1993). *Science in Russia and the Soviet Union: A Short History*. Cambridge: Cambridge University Press.

Graham, L. (2015). A rise in nationalism in Putin's Russia threatens the country's science – again. *The Conversation,* accessed at https://theconversation.com/a-rise-in-nationalism-in-putins-russia-threatens-the-countrys-science-again-41403.

Grant, M. (1916). *The Passing of the Great Race: Or, the Racial Basis of European History*. New York: Charles Scribner's Sons.

Gravlee, C., Bernard, H. and Leonard, W. (2003). Heredity, environment, and cranial form: a reanalysis of Boas's immigrant data. *American Anthropologist*, 105(1), pp. 125–138.

Greene, J. (1981). *Science, Ideology, and World View: Essays in the History of Evolutionary Ideas*. Berkeley: University of California Press.

Greene, J. (1990). The interaction of science and world view in Sir Julian Huxley's evolutionary biology. *Journal of the History of Biology*, 23(1), pp. 39–55.

Greene, J. (1992). From Aristotle to Darwin: reflections on Ernst Mayr's interpretation in the growth of biological thought. *Journal of the History of Biology*, 25(2), pp. 257–284.

Griesemer, J. (2002). What is "epi" about epigenetics? *Annals of the New York Academy of Sciences*, 981(1), pp. 97–110.

Griesemer, J. and Wimsatt, W. (1989). Picturing Weismannism: a case study of conceptual evolution. In: Ruse, M., ed. *What the Philosophy of Biology Is: Essays Dedicated to David Hull*. Dordrecht: Kluwer Academic Publishers, pp. 75–137.

Griffin, R. (2007). Tunnel visions and mysterious trees: modernist projects of national and racial regeneration, 1880–1939. In: Turda, M. and Weindling, P., eds. *Blood and Homeland Eugenics and Racial Nationalism in Central and Southeast Europe, 1900–1940*. Budapest, New York: Central European University Press.

Griffiths, P. (2004). Instinct in the '50s: the British reception of Konrad Lorenz's theory of instinctive behaviour. *Biology and Philosophy*, 19(4), pp. 609–631.

Griffiths, P., and Stotz K. (2013). *Genetics and Philosophy*. Cambridge: Cambridge University Press.

Gross, A. (1996). *The Rhetoric of Science*. Cambridge & London: Harvard University Press.

Grossniklaus, U., Kelly, W., Ferguson-Smith, A., Pembrey, M. and Lindquist, S. (2013). Transgenerational epigenetic inheritance: how important is it? *Nat Rev Genet*, 14(11), pp. 820–820.

Gudding, G. (1996). The phenotype/genotype distinction and the disappearance of the body. *Journal of the History of Ideas*, 57(3), pp. 525–545.

Guthman, J., and Mansfield, B. (2013). The implications of environmental epigenetics: a new direction for geographic inquiry on health, space, and nature-society relations. *Progress in Human Geograph*, 37(4), pp. 486–504.

Hacking, I. (1983). *Representing and Intervening: Introductory Topics in the Philosophy of Natural Science*. Cambridge: Cambridge University Press.

Hacking, I. (2002). *Historical Ontology*. Cambridge, Mass.: Harvard University Press.

Haig, D. (1993). Genetic conflicts in human pregnancy. *The Quarterly Review of Biology*, 68(4), p. 495.

Haig, D. (2011). Commentary: The epidemiology of epigenetics. *International Journal of Epidemiology*, 41(1), pp. 13–16.

Haldane, J. (1924). *Daedalus: Or, Science and the Future*. New York: E.P. Dutton & Company.

Haldane, J. (1932). *The Inequality of Man*. London: Chatto & Windus.

Haldane, J. (1938). *Heredity and Politics*. New York: Norton.

Haldane, J. (1944). Reshaping plants and animals. In: Huxley, J., Haldane, J., Drummond, J., Crick, W., Witts, L., Mackintosh, J. and Appleton, E., eds. *Reshaping man's heritage*. London: George Allen & Unwin, pp. 31–38.

Hale, P. (2009). Of mice and men: evolution and the socialist utopia. William Morris, H. G. Wells, and George Bernard Shaw. *Journal of the History of Biology*, 43(1), pp 17–66.

Haller, J. (1971). *Outcasts from Evolution: Scientific Attitudes of Racial Inferiority, 1859–1900*. Urbana, IL: University of Illinois Press.

Haller, M. (1963). *Eugenics: Hereditarian Attitudes in American Thought*. New Brunswick, New Jersey: Rutgers University Press.

Hallgrímsson, B. and Hall, B., eds. (2011). *Epigenetics: Linking Genotype and Phenotype in Development and Evolution*. Berkeley, CA: University of California Press.

Hamilton, W. D. (1964). The genetical evolution of social behavior I and II. *Journal of Theoretical Biology*, 7, pp. 1–52.

Hanson, M. (2015). The birth and future health of DOHaD. *Journal of Developmental Origins of Health and Disease*, pp. 1–4.

Hanson, M., Low F., and Gluckman, P. (2011). Epigenetic epidemiology: the rebirth of soft inheritance, *Ann Nutr Metab*, 58(Suppl. 2), pp. 8–15.

Haraway, D. (1976). *Crystals, Fabrics, and Fields: Metaphors that Shape Embryos.* Berkeley CA: North Atlantic Books.

Haraway, D. (1988). Situated knowledges: the science question in feminism and the privilege of partial perspective. *Feminist Studies*, 14, pp. 575–599.

Haraway, D. (1990). *Primate Visions: Gender, Race, and Nature in the World of Modern Science.* New York: Routledge.

Harding, S. (1986). *The Science Question in Feminism.* Ithaca: New York Cornell.

Harding, S. (1991). *Whose Science? Whose Knowledge?* Ithaca: New York Cornell.

Harding, S. (2006). *Science and Social Inequality: Feminist and Postcolonial Issues.* Urbana and Chicago: University of Illinois Press.

Harding, S. (2013). (Ed.). *The Postcolonial Science and Technology Studies Reader.* Durham and London: Duke University Press.

Harwood, J. (1986). Ludwik Fleck and the Sociology of Knowledge. *Social Studies of Science*, 16, pp. 173–187.

Harwood, J. (1993). *Styles of Scientific Thought: The German Genetics Community 1900–1933.* Chicago: University of Chicago Press.

Hazard A. Q. (2012), *Postwar Anti-racism: The United States, UNESCO, and Race, 1945–1968.* New York: Palgrave Macmillan.

Heard, E. and Martienssen, R. (2014). Transgenerational epigenetic inheritance: myths and mechanisms. *Cell*, 157(1), pp. 95–109.

Heckman, J. (2012). 'Early Childhood Development: Learning, Behavior and Health' NICHD Colloquium 5 December Bethesda, Maryland.

Hedlund, M. (2012). Epigenetic responsibility. *Medicine Studies*, 3(3), pp. 171–183.

Heijmans, B., Tobi, E., Stein, A., Putter, H., Blauw, G., Susser, E., Slagboom, P., and Lumey, L. (2008). Persistent epigenetic differences associated with prenatal exposure to famine in humans. *Proceedings of the National Academy of Sciences USA*, 105, pp. 17046–17049.

Heijmans, B., Tobi, E., Lumey T. H. (2009). The epigenome: archive of the prenatal environment. *Epigenetics*, 4(8), pp. 526–531.

Heijmans B.T. and Mill J. (2012). Commentary: the seven plagues of epigenetic epidemiology. *Int J Epidemiol*, 41(1), pp. 74–78.

Hennessy, M. (2012). Odds stacked against Glasgow's poorest – even before birth. *The Irish Times*, 27 Jan.

Herbert, S. (1910). Eugenics and socialism, *Eugenics Review*, 2(2), pp. 116–123.

Herceg, Z. and Murr, R. (2010). Mechanisms of histone modifications. In: Tollefsbol, T., ed. *Handbook of Epigenetics: The New Molecular and Medical Genetics*. London: Academic, pp. 25–45.

Hertzler, J. (1923). *The History of Utopian Thought.* London: G. Allen & Unwin.

Hey, J. (2011). Regarding the confusion between the population concept and Mayr's "population thinking". *Q Rev Biol.*, 86(4), pp. 253–264.

Hobsbawm, E. (1995), *The Age of Extremes: The Short Twentieth Century, 1914–1991*, Abacus, London.

Hodge, M. (1985). Darwin as a lifelong generation theorist. In: Kohn, D., ed. *The Darwinian Heritage.* Princeton, NJ: Princeton University Press.

Hodson, K. (1968). The eugenics review 1909–1968, *Eugenics Review.*

Hofstadter, R. (1944). *Social Darwinism in American Thought, 1860–1915.* University of Pennsylvania Press, Philadelphia, PA.

Holliday, R. (1990). DNA methylation and epigenetic inheritance. *Philosophical Transactions of the Royal Society of London Series B, Biological Sciences*, 326(1235), pp. 329–338.
Holliday, R. (2006). Epigenetics: a historical overview. *Epigenetics*, 1(2), pp. 76–80.
Hollinger D. (1998). *Science Jews and Secular Culture: Studies in Mid-Twentieth-Century American Intellectual History*, Princeton, N.J.: Princeton University.
Holton, G. (1978). *The Scientific Imagination*. Cambridge: Cambridge UP.
Hubbard, R. (1990). *The Politics of Women's Biology*. New Brunswick, NJ: Rutgers University Press.
Hughes, A. (1959). *A History of Cytology*. London: Abelard-Schuman.
Huxley, A. (1931). *What Dare I Think? The Challenge of Modern Science to Human Action and Belief*. London: Chatto and Windus.
Huxley, A. (1932). *Brave New World*. London: Chatto & Windus.
Huxley, J. (1934). *If I Were Dictator*. New York: Harper & Bros.
Huxley, J. (1936). Galton lecture: eugenics and society. In: Huxley, J., ed. (1941). *The Uniqueness of Man*. London: Chatto and Windus.
Huxley, J. (1942). *Evolution: The Modern Synthesis*. New York: Harper & Brothers.
Huxley, J. (1947). *UNESCO: Its Purpose and Philosophy*. Washington, D.C.: Public Affairs Press.
Huxley, J. (1949). *Soviet Genetics and World Science*. London: Chatto & Windus.
Huxley, J. (1955). Guest editorial: evolution, cultural and biological. *Yearbook of Anthropology*, pp. 2–25.
Huxley, J. (1957). "Transhumanism" in *New Bottles for New Wine*, London: Chatto & Windus, 1957; pp. 13–17.
Huxley, J. (1962). Eugenics in Evolutionary Perspective. *Eugen Rev.*, 54(3), pp. 123–141.
Huxley, J., Haddon, A. with a contribution of Carr-Saunders, A. (1936). *We Europeans: A Survey of "Racial" Problems*. New York: Harper.
Inge, W. (1909). Some moral aspects of eugenics. *Eugenics Review*, 1(1), p. 26.
Ingold, T. and Palsson, G. (2013). *Biosocial Becomings: Integrating Social and Biological Anthropology*. Cambridge: Cambridge University Press.
Iriye, A. (Ed.) (2014). *Global Interdependence The World after 1945*. Cambridge, MA: Harvard UP.
Jablonka, E., and Lamb, M. (1995). *Epigenetic Inheritance and Evolution: The Lamarckian Dimension*. Oxford: Oxford University Press.
Jablonka, E., and Lamb, M. (2005). *Evolution in Four Dimensions*. Cambridge, Mass.: MIT Press.
Jablonka, E., and Lamb, M. (2008). Soft inheritance: challenging the modern synthesis. *Genetics and Molecular Biology*, 31(2), pp. 389–395.
Jablonka, E., and Lamb, M. (2014). *Evolution in Four Dimensions*, 2nd ed. Cambridge, Mass: MIT Press.
Jablonka, E., and Raz, G. (2009). Transgenerational epigenetic inheritance: prevalence, mechanisms, and implications for the study of heredity and evolution. *Quarterly Review of Biology*, 84, pp. 131–176.
Jackson, J. (2001). "In ways unacademical": the reception of Carleton S. Coon's the origin of races. *Journal of the History of Biology*, 34(2), pp. 247–285.
Jackson, J. (2005). *Science for Segregation: Race, Law, and the Case against Brown V. Board of Education*. New York: New York University Press.

Jackson, J. and Weidman, N. (2004). *Race, Racism, and Science: Social Impact and Interaction*. Santa Barbara, Calif.: ABC-CLIO.
Jackson, J. and Weidman, N., eds. (2005). *Race, Racism, and Science: Social Impact and Interaction*. New Brunswick, NJ: Rutgers University Press.
Jacob, F. (1973). *The Logic of Life: A History of Heredity*. New York: Pantheon. 2nd ed. (1993). Princeton, NJ: Princeton University Press.
Jasanoff, S., ed. (2004). *States of Knowledge. The Co-production of Science and the Social Order*. New York: Routledge.
Jasanoff, S. (2005). *Designs on Nature: Science and Democracy in Europe and the United States*. Princeton NJ: Princeton University Press.
Jasanoff, S. (2012). *Science and Public Reason*. New York: Routledge.
Jasienska, G. (2009). Low birth weight of contemporary African Americans: an intergenerational effect of slavery? *Am J Human Biology*, 21, pp. 16–24.
Jirtle, R. (2012). Epigenetics: how genes and environment interact. Lecture, delivered at the *NIH director's Wednesday afternoon lecture series*. April 18, 2012. Accessed November 1, 2013.
Johannsen, W. (1909). *Elemente der exakten erblichkeitslehre*. Jena: Gustav Fischer.
Johannsen, W. (1911). The genotype conception of heredity. *The American Naturalist*, 45(531), p. 129.
Johannsen, W. (1923). Some remarks about units in heredity. *Hereditas*, 4, pp. 133–141.
Johnson, R. and Popenoe, P. (1918). *Applied Eugenics*. New York: Macmillan Company.
Johnston, T. (1995). The influence of Weismann's germ-plasm theory on the distinction between learned and innate behavior. *J. Hist. Behav. Sci.*, 31(2), pp. 115–128.
Jones, G. (1982). Eugenics and social policy between the wars. *The Historical Journal*, 25(3). pp. 717–728.
Jones, G. (1986). *Social Hygiene in Twentieth Century Britain*. London: Croom Helm.
Jordanova, L. J. (1984). *Lamarck*. Oxford: Oxford University Press.
Judson. (1996). *The Eighth Day of Creation: Makers of the Revolution in Biology* (1979). Touchstone Books, Cold Spring Harbor Laboratory Press
Jumonville, N. (2002). The cultural politics of the sociobiology debate. *Journal of the History of Biology*, 35, pp. 569–593.
Kaati, G., Bygren, L. O., Edvinsson, S. (2002). Cardiovascular and diabetes mortality determined by nutrition during parents' and grandparents' slow growth period. *Eur J Hum Genet*, 10, pp. 682–688.
Kahn, J. (2013). *Race in a Bottle: the Story of BiDil and Racialized Medicine in a Post-Genomic Age*. New York: Columbia University Press.
Kalmus, H. (1983). The scholastic origins of Mendel's concepts. *History of Science*, 21, pp. 61–83.
Kammerer, P. (1912). Körper kultur und rasse. *Das oesterreichische Sanitatswesen*, 24, pp. 441–452.
Kammerer, P. (1920). *Das biologische Zeitalter*. Hamburg: Hamburger Verlag.
Kammerer, P. (1924). *The Inheritance of Acquired Characteristics*. New York: Boni and Liveright.
Kammerer, P. and Steinach, E. (1923). *Rejuvenation and the Prolongation of Human Efficiency: Experiences with the Steinach-Operation on Man and Animals*. New York: Boni and Liveright.

Kappeler, L. and Meaney, M. (2010). Epigenetics and parental effects. *Bioessays*, 32, pp. 818–827.

Katz, M. (2013). The biological inferiority of the undeserving poor. [online] *Social Work and Society International Online Journal*, 11(1). Available at http://www.socwork.net/sws/article/view/359/709. Accessed November, 2014.

Kaufman, J. and Hall, S. (2003). The slavery hypertension hypothesis: dissemination and appeal of a modern race theory. *Epidemiology*, 14(1), pp. 111–118.

Kay, L. (1993). *The Molecular Vision of Life*, Caltech, the Rockefeller Foundation, and the Rise of the New Biology.

Keller, E. F. (1983). *A Feeling for the Organism: The Life and Work of Barbara McClintock*. New York: Freeman.

Keller, E. F. (1985). *Reflections on Gender and Science*. New Heaven: Yale University Press.

Keller, E. F. (1988). Demarcating public from private values in evolutionary discourse. *J Hist Biol*, 21(2), pp. 195–211.

Keller, E. F. (1992). Between language and science: the question of directed mutation in molecular genetics. *Perspectives in Biology and Medicine*, 35(2), pp. 292–306.

Keller, E. F. (1995). *Refiguring Life: Metaphors of Twentieth Century Biology*. New York: Columbia University Press.

Keller, E. F. (1996). *Refiguring Life: Metaphors of Twentieth-Century Biology*. New York: Columbia University Press.

Keller, E. F. (2000). *The Century of the Gene*. Cambridge, Mass.: Harvard University Press.

Keller, E. F. (2001). Making a Difference: Feminist Movement and Feminist critiques of Science. In: Creager, A., Lunbeck, E and Schiebinger, L, eds. *Feminism in Twentieth Century Science, Technology, and Medicine*. Chicago: University of Chicago Press.

Keller, E. F. (2010). *The Mirage of a Space Between Nature and Nurture*. Durham, NC: Duke University Press.

Keller, E. F. (2011). Genes, genomes, and genomics. *Biol Theory*, 6, pp. 132–140.

Keller, E. F. (2014). From gene action to reactive genomes. *J Physiol*, 592(11), pp. 2423–2429.

Keller, E. F. (2015). The postgenomic genome. In: Richardson, S. and Stevens, H., eds. *Postgenomics: Perspectives on Biology after the Genome*. Durham, NC: Duke University Press, pp. 9–31.

Kelly, A (1981). *The Descent of Darwin: The Popularization of* Darwinism *in Germany: 1860–1914*. Chapel. Hill: University of North Carolina.

Kevles, D. (1977). The national science foundation and the debate over postwar research policy, 1942–45. *Isis*, 68, pp. 5–26.

Kevles, D. (1985). *In the Name of Eugenics*. Cambridge, Mass: Harvard University Press.

Knorr-Cetina, K. D. (1981). *The Manufacture of Knowledge*. Pergamon Press, Oxford.

Knorr-Cetina, K. D. (1983). Towards a Constructivist Interpretation of Science. In: Knorr-Cetina, K. D. and Mulkay, M., eds., Science Observed, Sage, CA.

Koch, L. (2004). The meaning of eugenics: reflections on the government of genetic knowledge in the past and present. *Science in Context*, 17, pp. 315–331.

Koenig, B, Soo-Jin Lee, S. Richardson, S., eds. (2008). *Revisiting Race in a Genomic Age*. New Brunswick, NJ: Rutgers University Press.

Koestler, A. (1971). *The Case of the Midwife Toad*. London: Hutchinson.

Koonin, E. (2012). *The Logic of Chance: The Nature and Origin of Biological Evolution*. Upper Saddle River, NJ: Pearson Education.

Krementsov, N. (1997). *Stalinist Science*. Princeton, NJ: Princeton University Press.

Krementsov, N. (2010). Eugenics in Russia and the Soviet Union. In: Bashford, A. and Levine, P., eds. *The Oxford handbook of the history of eugenics*. New York: Oxford University Press, pp. 413–429.

Krementsov, N. (2011). From "beastly philosophy" to medical genetics: eugenics in Russia and the Soviet Union. *Annals of Science*, 68(1), pp. 61–92.

Krimbas, C. (1994). The evolutionary worldview of Theodosius Dobzhansky. In: Adams, M. B., ed. *The Evolution of Theodosius Dobzhansky. Essays on His Life and Thought in Russia and America*. Princeton, NJ: Princeton University Press, pp. 179–194.

Kroeber, A. (1915). Eighteen professions. *American Anthropologist*, 17(2), pp. 283–288.

Kroeber, A. (1916a). The cause of the belief in use inheritance. *The American Naturalist*, 50(594), p. 367.

Kroeber, A. (1916b). Inheritance by magic. *American Anthropologist*, 18(1), pp. 19–40.

Kroeber, A. (1917). The superorganic. *American Anthropologist*, 19(2), pp. 163–213.

Kroeber, A. (1952). *The Nature of Culture*. Chicago: University of Chicago Press.

Kroeber, A. (1953). *Anthropology Today; An Encyclopedic Inventory*. Chicago: Univ. of Chicago Press.

Kroeber, T. (1970). *Alfred Kroeber; A Personal Configuration*. Berkeley CA: University of California Press.

Krönfeldner, M. (2009). "If there is nothing beyond the organic..." heredity and culture at the boundaries of anthropology in the work of Alfred L. Kroeber's N TM. *Zeitschrift fiir Geschichte der Wissenschaften, Technik und Medizin*, 17, pp. 107–133.

Krüger, A. (1998). A horse breeder's perspective: scientific racism in Germany, 1870–1933. In: Finzsch, N. and Schirmer, D., eds. (1998). *Identity and Intolerance: Nationalism, Racism, and Xenophobia in Germany and the United States*. Washington, D.C.: German Historical Institute, pp. 371–395.

Kucharski, R., Maleszka, J., Foret, S., and Maleszka, R. (2008). Nutritional control of reproductive status in honey bees via DNA methylation. *Science*, 319, pp. 1827–1830.

Kühl, S. (1994). *The Nazi Connection: Eugenics, American Racism, and German National Socialism*. New York: Oxford University Press.

Kühl, S. (2013). *For the Betterment of Race*. London: Palgrave.

Kuhn, T. (2000). *The Road Since Structure*, Conant, James and Haugeland, John, eds. Chicago: University of Chicago Press.

Kuzawa, C. (2004). The fetal origins of developmental plasticity: Life history, adaptation, and disease. *American Journal of Human Biology*, 16(2), pp. 232–236.

Kuzawa, C. (2005). Fetal origins of developmental plasticity: are fetal cues reliable predictors of future nutritional environments? *Am J Hum Biol*, 17, pp. 5–21.

Kuzawa, C. (2008). The developmental origins of adult health: intergenerational inertia in adaptation and disease. In: Trevathan, W., Smith E., and McKenna J., eds. *Evolutionary Medicine and Health: New Perspectives*. Oxford: Oxford University Press, pp. 325–349.

Kuzawa, C. and Bragg (2011). Plasticity in human life history strategy: implications for contemporary human variation and the evolution of genus homo, *Current Anthropology*, *53*(6), pp. 369–385.

Kuzawa, C. and Bragg, J. (2012). Plasticity in human life history strategy: implications for contemporary human variation and the evolution of genus homo. *Current Anthropology*, 53(Suppl. 6), pp. s369–s382.

Kuzawa, C. and Quinn, E. (2009). Developmental origins of adult function and health: Evolutionary hypotheses. *Annual Review of Anthropology*, 38, pp. 131–147.

Kuzawa, C. and Sweet, E. (2009). Epigenetics and the embodiment of race: Developmental origins of US racial disparities in cardiovascular health. *American Journal of Human Biology*, 21(1), pp. 2–15.

Kuzawa, C. and Thayer, Z. M. (2011). Timescales of human adaptation: the role of epigenetic processes, *Epigenomics*, 3, 2011, pp. 221–234.

Lallement, M. (1993). *Histoire des idées sociologique: des origines à Weber*. 4th ed. Paris: Armand Colin.

Lamarck, J. B. (1809/2011). *Zoological philosophy: an exposition with regard to the natural history of animals with introductory essays by David L. Hull and Richard W. Burkhardt Jr*. New York: Cambridge University Press.

Lamb, M. (2011). In: Gissis, S. and Jablonka, E., eds. *Transformations of Lamarckism: From Subtle Fluids to Molecular Biology*. Cambridge, Mass.: MIT Press.

Lamont, M. and Molnár, V. (2002). The study of boundaries in the social sciences. *Annual Review of Sociology*, 28(1), pp. 167–195.

Landecker, H. (2011). Food as exposure: nutritional epigenetics and the new metabolism. *BioSocieties*, 6(2), pp. 167–194.

Landecker, H. (2016). The social as signal in the body of chromatin. In: Meloni, M., Williams, S., Martin, P. eds. *Biosocial Matters: Rethinking Sociology-Biology Relations in the Twenty-First Century*. Oxford: Wiley-Blackwell.

Landecker, H. and Panofsky, A. (2013). From social structure to gene regulation, and back: a critical introduction to environmental epigenetics for sociology. *Annual Review of Sociology*, 39, pp. 333–357.

Laporte, L. (2000). *George Gaylord Simpson. Paleontologist and Evolutionist*. New York: Columbia.

Lappé, M. and Landecker, H. (2015). How the genome got a life span, new genetics and society. *New Genetics and Society*, 34(2), pp. 152–176.

Largent, M. (2008). *Breeding Contempt: The History of Coerced Sterilization in the United States*, New Brunswick, NJ: Rutgers University Press.

Larson, E. J. (1991). The rhetoric of eugenics: expert authority and the mental deficiency bill. *British Journal of the History of Science*, 24, pp. 45–60.

Latour, B. (1988). *Science in Action. How to Follow Scientists and Engineers through Society*. Cambridge, Mass: Harvard University Press.

Latour, B. (1993). *We Have Never Been Modern*. trans. C. Porter. Cambridge, Mass.: Harvard University Press.

Latour, B. (2009). *Politics of Nature. How to Bring the Sciences Into Democracy*. Cambridge, Mass: Harvard University Press.

Laublicher, M. and Maienschein J., eds. (2007). *From Embryology to Evo-Devo: A History of Developmental Evolution*. Cambridge, Mass: MIT Press.

Laudan, L. (1977). *Progress and its Problems*, Berkeley: University of California Press.

Laughlin, H. H. (1919). The relation of eugenics to other sciences. *The Eugenics Review*, 11(2), pp. 53–65.

Lecourt, D. (1969). *L'épistémologie historique de Gaston Bachelard*. Paris: Vrin.

Lederberg, J. and McCray, A. (2001). Ome sweet omics: a genealogical treasure of words. *Scientist*, 15(7), p. 8.

Lemke, T. (2011). *Biopolitics: An Advanced Introduction*. New York, NY: New York UP.

Lenoir, T. ed. (1998). *Inscribing Science: Scientific Texts and the Materiality of Communication*. Stanford: Stanford University Press.

Lenoir, T. (2010). Epistemology Historicized. In Rheinberger H. J., ed. *An Epistemology of the Concrete: Twentieth-Century Histories of Life*.

Leonard, T. (2005). Retrospectives: eugenics and economics in the progressive era. *Journal of Economic Perspectives*, 19(4), pp. 207–224.

Leonard, T. C. (2009). Origins of the myth of social Darwinism: the ambiguous legacy of Richard Hofstadter's Social Darwinism in American Thought. *Journal of Economic Behavior & Organization*, 71(1), pp. 37–51.

Lerner, I. M. (1954). *Genetic Homeostasis*. Edinburgh, UK: Oliver and Boyd.

Lerner, R. M. (1992). *Final Solutions: Biology, Prejudice, and Genocide*. University Park, PA: Pennsylvania State University Press.

Leroux, R. (1998). *Histoire et sociologie en France: de l'histoire-science à la sociologie durkheimienne*. Paris: Presses universitaires de France.

Levine, L., ed. (1995). *Genetics of Natural Populations: The Continuing Importance of Theodosius Dobzhansky*. New York: Columbia University Press.

Levine, P. (2010). Anthropology, colonialism and eugenics. In: *The Oxford Handbook of the History of Eugenics*. New York: Oxford University Press, pp. 43–61.

Levy, D. and Sznaider, N. (2002). Memory unbound: the Holocaust and the formation of cosmopolitan memory. *European Journal of Social Theory*, 5(1), pp. 87–106.

Levy, D. and Sznaider, N. (2004). The institutionalization of cosmopolitan morality: the Holocaust and human rights. *Journal of Human Rights*, 3(2), pp. 143–157.

Lewontin, R. (1974). *The Genetic Basis of Evolutionary Change*. New York: Columbia University Press.

Lewontin, R. (1983). Gene, organism, and environment. In: Bendall, D., ed. *Evolution from Men to Molecules*. Cambridge: Cambridge University Press, pp. 273–285.

Lewontin, R. (1997). Dobzhansky's Genetics and the origin of Species: Is It Still Relevant? *Genetics* 147, 2, pp. 351–355.

Lewontin, R. (2011). It's even less in your genes. *New York Review of Books*, May 26.

Leykin I. (2015). Rodologia: Genealogy as Therapy in Post-Soviet Russia. *Ethos*, 43(2), pp. 135–164.

Lifton, R. J. (1986). *The Nazi Doctors: Medical Killing and the Psychology of Genocide*. New York: Basic Books.

Liftton, R. J. (2000). *The Nazi Doctors: Medical Killing and the Psychology of Genocide.* New York: Basic Books.

Lim, J. and Brunet, A. (2013). Bridging the transgenerational gap with epigenetic memory. *Trends in Genetics*, 29(3), pp. 176–186.

Linke, U. (1999). *Blood and Nation: The European Aesthetics of Race.* Philadelphia: University of Pennsylvania Press.

Lipphardt, V. (2012). Isolates and crosses in human population genetics; or, a contextualization of German race science. *Current Anthropology*, 53(S5), pp. S69–S82.

Lipphardt, V. (2014). Geographical distribution patterns of various genes: genetic studies of human variation after 1945. *Studies in History and Philosophy of Science Part C: Studies in History and Philosophy of Biological and Biomedical Sciences*, 47, pp. 50–61.

Little, M. A. (2012). Human Population Biology in the Second Half of the Twentieth Century. *Current Anthropology*, 53, pp. S126–S138.

Little, M. A. and Sussman, R. (2010). History of Biological Anthropology. In: Larsen C. S., ed. *A Blackwell Companion to Biological Anthropology*, Wiley-Blackwell.

Lock, M. (1993). *Encounters with Aging: Mythologies of Menopause in Japan and North America.* Berkeley: University of California Press.

Lock, M. (2001). Containing the elusive body. *The Hedgehog Review*, 3(2), pp. 65–78.

Lock, M. (2005). Eclipse of the gene and the return of divination. *Current Anthropology*, 46, pp. 47–70.

Lock, M. (2012). The epigenome and nature/nurture reunification: a challenge for anthropology. *Medical Anthropology*, 32(4), pp. 291–308.

Lock, M. (2013). The lure of the epigenome. *The Lancet*, 381, pp. 1896–1897.

Lock, M. (2015). Comprehending the body in the era of the epigenome. *Current Anthropology*, 56(2), pp. 151–177.

Lock, M., and Nguyen, V. (2010). *An Anthropology of Biomedicine.* Oxford: Wiley-Blackwell.

Loi, M., Del Savio, L., and Stupka, E. (2013). Social epigenetics and equality of opportunity. *Public Heath Ethics*, 6, pp. 142–153.

Logan, C. (2013). *Hormones, Heredity, and Race: Spectacular Failure in Interwar Vienna.* New Brunswick, NJ: Rutgers University Press.

Logan, C. and Johnston, T. (2007). Synthesis and separation in the history of "nature" and "nurture". *Developmental Psychobiology*, 49(8), pp. 758–769.

Lomax, E. (1977). Hereditary or acquired disease? Early nineteenth century debates on the cause of infantile scrofula and tuberculosis. *Journal of the History of Medicine and Allied Sciences*, 32(4), pp. 356–374.

Lomax, E. (1979). Infantile syphilis as an example of nineteenth century belief in the inheritance of acquired characteristics. *Journal of the History of Medicine and Allied Sciences*, 34(1), pp. 23–39.

Lombardo, P. (2008). *Three Generations, No Imbeciles: Eugenics, the Supreme Court, and Buck V. Bell.* Baltimore, MD: Johns Hopkins University Press.

Longino, Helen E. (1990). *Science as Social Knowledge: Values and Objectivity in Scientific Inquiry*, Princeton: Princeton University Press.

Longino H. (1996). Cognitive and non-cognitive values in science: Rethinking the Dichotomy. In: Nelson, L. H. and Nelson, J., eds. *Feminism and Philosophy of Science*, Dordrecht: Kluwer.

López-Beltrán, C. (2004). In the cradle of heredity; French physicians and l'hérédité naturelle in the early 19th century. *Journal of the History of Biology*, 37(1), pp. 39–72.

López-Beltrán, C. (2007). The medical origins of heredity. In: Heredity Produced, Staffan Müller-Wille and Hans-Jörg Rheinberger. Cambridge, MA: MIT Press, pp. 105–132.

Ludmerer, K. (1972). *Genetics and American Society: A Historical Appraisal*. Baltimore, MD: Johns Hopkins University Press.

Lutz, P. and Turecki, G. (2014). DNA methylation and childhood maltreatment: from animal models to human studies. *Neuroscience*, 264, pp. 142–156.

MacBride, E. (1923). Heredity and eugenics. *Eugenics Review*, 15(3), pp. 508–510.

MacBride, E. (1924) *An Introduction to the Study of Heredity*. New York: H. Holt & Co.

MacBride, E. (1926). The dogma of evolution. *Eugenics Review*, 18(1), pp. 38–41.

MacBride, E. (1930a). Eugenic sterilization. *Nature*, 126, pp. 301–302.

MacBride, E. (1930b). Sterilization as a practical eugenic policy. *Nature*, 125, pp. 40–42.

MacBride, E. (1931). Birth control and human biology. *Nature*, 127, pp. 509–511.

MacBride, E. (1936). Cultivation of the unfit. *Nature*, 137, pp. 44–45.

MacBride, E. W. (1937a). The Spanish Civil War. *The Times*, 22 July.

MacBride, E. W. (1937b) Letter to the Editor. *The Times*, 30 July.

Mackenzie, D. (1979). Karl Pearson and the professional middle class. *Annals of Science*, 36(22), pp. 124–144.

Mackenzie, D. (1982). *Statistics in Britain, 1865–1930: The Social Construction of Scientific Knowledge*. Edinburgh: Edinburgh University Press.

Mackenzie, A. (2015). Machine learning and genomic dimensionality: from features to landscapes in postgenomics. In: Richardson, S. and Stevens, H., eds. *Postgenomics: perspectives on biology after the genome*. Durham, NC: Duke University Press, pp. 73–102.

Maderspacher, F. (2010). Lysenko rising. *Current Biology*, 20, pp. 835–837.

Maestripieri, D. and Mateo, J., eds. (2009). *Maternal Effects in Mammals*. Chicago: University of Chicago Press.

Maher, B. (2008). Personal genomes: The case of the missing heritability. *Nature*, 456(7218), pp. 18–21.

Maienschein, J. (1991). *Transforming Traditions in American Biology, 1880–1915*. Baltimore: The John Hopkins University Press.

Maienschein, J. (2005). *Whose View of Life? Embryos, Cloning, and Stem Cells*. Cambridge, Mass.: Harvard University Press.

Mannheim, K. (1953). *Essays in Sociology and Social Psychology*. London: Routledge.

Mansfield, B. (2012). Race and the new epigenetic biopolitics of environmental health. *BioSocieties*, 7(4), pp. 352–372.

Mansfield, B. and Guthman, J. (2015). Epigenetic life: biological plasticity, abnormality, and new configurations of race and reproduction. *Cultural Geographies*, 22(1), pp. 3–20.

Marcel, J. and Guillo, D. (2006). Durkheimian sociology, biology and the theory of social conflict. *International Social Science Journal*, 58, pp. 83–100.

Marks, J. (2008). Scientific racism, history of. In: Moore, J., ed. *Encyclopedia of Race and Racism*, vol. 3. Detroit: Macmillan Reference USA, pp. 1–16.

Marks, J. (2010) The two twentieth century crises of Racial Anthropology. In: M. A. Little, K. A. R. Kennedy, eds. *Histories of American Physical Anthropology in the Twentieth Century*. Lanham, Maryland: Lexington Books, pp. 187–206.

Marks, J. (2012). The Origins of Anthropological Genetics. *Current Anthropology*, 53(S5), pp. S161–S172.

Marmot, M. (2005). Social determinants of health inequalities, *The Lancet*, 365(9464), pp. 1099–1104.

Marmot, M., Smith, G., Stansfeld, S., Patel, C., North, F., Head, J., White, I., Brunner, E., and Feeney, A. (1991). Health inequalities among British civil servants: the Whitehall II study. *The Lancet*, 337, pp. 1387–1393.

Martin E. (1991). The egg and the sperm: how science has constructed a romance based on stereotypical male-female roles. *Signs*, 16, pp. 4485–4501.

Massey, D. (2013). Inheritance of poverty or inheritance of place? The emerging consensus on neighborhoods and stratification. *Contemporary Sociology: A Journal of Reviews September*, 42(5), pp. 690–695.

Mattick, J. (2003). Challenging the dogma: the hidden layer of non-protein-coding RNAs in complex organisms. *Bioessays*, 25(10), pp. 930–939.

Mattick, J. (2004). Opinion: RNA regulation: a new genetics?. *Nat Rev Genet*, 5(4), pp. 316–323.

Maxmen, A. (2012). Researchers highlight the impact of slavery on health and disease. [online] *Nature News Blog*. Available at http://blogs.nature.com/news/2012/07/researchers-highlight-the-impact-of-slavery-on-health-and-disease.html. Accessed 24 June, 2015.

Maynard Smith, J. (1968). Eugenics and Utopia, *Daedalus*, 117, pp. 73–92.

Mayr E. (1942). *Systematics and the Origin of Species from the Viewpoint of a Zoologist* New York: Columbia University Press.

Mayr, E. (1959). Darwin and the evolutionary theory in biology. In: Meggars, B., ed. *Evolution and Anthropology: A Centennial Appraisal*. Washington, D.C.: Anthropological Society of Washington, pp. 1–10.

Mayr, E. (1962). Origin of the human races [Review of *The origin of races* by C. S. Coon]. Science, 138, pp. 420–422.

Mayr, E. (1963). *Animal Species and Evolution*. London: Oxford University Press.

Mayr, E. (1980). Prologue: some thoughts on the history of the evolutionary synthesis: In Mayr, E., and W. B. Provine, eds. *The Evolutionary Synthesis: Perspectives on the Unification of Biology*. Cambridge, MA: Harvard University Press.

Mayr, E. (1982). *The Growth of Biological Thought: Diversity, Evolution, and Inheritance*. Cambridge, Mass.: Belknap Press.

Mayr, E. (1988). *Toward a New Philosophy of Biology: Observations of an Evolutionist*. Cambridge, Mass: Harvard UP.

Mayr, E. (1991). *One Long Argument: Charles Darwin and the Genesis of Modern Evolutionary Thought*. Cambridge, Mass.: Harvard University Press.

Mazumdar, P. (1992). *Eugenics, Human Genetics, and Human Failings: The Eugenics Society, Its Sources and Its Critics in Britain*. London: Routledge.

Mazumder et al. (2010). Lingering Prenatal Effects of the 1918 Influenza Pandemic on Cardiovascular Disease. *J Dev Orig Health Dis.*, 1(1), pp. 26–34.

McClintock, B. (1984). The significance of responses of the genome to challenge. *Science*, 226, pp. 792–801.

McDougall, W. (1927). An experiment for the testing of the hypothesis of Lamarck. *British Journal of Psychology*, 17(4), pp. 267–304.

McEwen, B., and Gianaros, P. (2010). Central role of the brain in stress and adaptation: links to socioeconomic status, health, and disease. *Annals of the New York Academy of Sciences*, 1186, pp. 190–222.

McGowan, P., Suderman, M., Sasaki, A., Huang, T., Hallett, M., Meaney, M. and Szyf, M. (2011). Broad epigenetic signature of maternal care in the brain of adult rats. *PLoS ONE*, 6, p. e14739.

McKee, J. (1993). *Sociology and the Race Problem: The Failure of a Perspective*. Urbana Champaign, Ill: University of Illinois Press.

McLaughlin, M. (2012). Babies born into poverty are damaged forever before birth. *The Scotsman* 24 January.

Meaney, M. (2001). Nature, nurture, and the disunity of knowledge. *Annals of the New York Academy of Sciences*, 935, pp. 50–61.

Meaney, M. (2010). Epigenetics and the biological definition of gene X environment interactions. Child Development, 81(1), pp. 41–79.

Meaney, M., and Szyf, M. (2005). Maternal care as a model for experience-dependent chromatin plasticity. *Trends in Neurosciences*, 28, pp. 456–463.

Medawar, P. B. (1960). The *Future* of *Man. The BBC Reith Lectures, 1959*. New York: Basic Books.

Meloni, M. (2010). Biopolitics for philosophers, *Economy and Society*, 39(4), pp. 551–566

Meloni, M. (2014). How biology became social, and what it means for social theory. *Sociol Rev*, 62(3), pp. 593–614.

Meloni, M. (2015a). Heredity 2.0: the epigenetics effect. *New Genetics and Society*, 34(2), pp. 117–124.

Meloni, M. (2015b). Epigenetics for the social sciences: justice, embodiment, and inheritance in the postgenomic age. *New Genetics and Society*, 34(2), pp. 125–151.

Meloni, M (2016) "From Boundary-Work to Boundary Object: How Biology Left and Re-Entered the Social Sciences" in *Biosocial Matters: Rethinking Sociology-Biology Relations in the Twenty-First Century*, edited by M. Meloni, S. Williams, P. Martin, Wiley-Blackwell, 2016.

Meloni, M. and Testa, G. (2014). Scrutinizing the epigenetics revolution. *BioSocieties*, 9(4), pp. 431–456.

Meloni. M., Williams, S., and P. Martin. eds. (2016). *Biosocial Matters: Rethinking Sociology-Biology Relations in the Twenty-First Century*. Wiley-Blackwell.

Merton, R. ([1938]1973]). Science and the social order. In: Storer, Norman, ed. *The Sociology of Science: Theoretical and Empirical Investigations*. Chicago: The University of Chicago Press, pp. 254–266.

Merton, R. (1942). A note on science and democracy. *Journal of Legal and Political Sociology*, 1, pp. 115–126. Reprinted as "Science and democratic social structure" in *Social Theory and Social Structure* ([1949, 1957]1968), pp. 604–615.

Mesoudi, A., Blanchet, S., Charmantier, A., Danchin, E., Fogarty, L., Jablonka, E., et. al. (2013). Is non-genetic inheritance just a proximate mechanism? A corroboration of the extended evolutionary synthesis. *Biol. Theory*, 7, pp. 189–195.

Métraux, A. (1950a). Race and civilization. *Courier*, 3(6–7), pp. 8–9.

Métraux, A. (1950b). UNESCO and race problems. *International Social Sciences Bulletin*, 2(3), p. 390.

Métraux, A. (1951). *American Anthropologist*, n.s. 53, p. 298.
Milam, E. (2010). *Looking for a Few Good Males: Female Choice in Evolutionary Biology*. Baltimore MD: Johns Hopkins University Press.
Mitman, G. (1988). From the population to society: the cooperative metaphors of W. C. Allee and A. E. Emerson. *Journal of the History of Biology*, 21(2), pp. 173–194.
Mitman, G. (1992). *The State of Nature: Ecology, Community, and American Social Thought, 1900–1950*. Chicago: The University of Chicago Press.
Montagu, M. F. A. (1951). *Statement on Race*. New York: Schuman.
Montagu, M. F. A. (1962). The Concept of Race. *American Anthropologist*, 64(5), pp. 919–928.
Moore, R. (2001). The "rediscovery" of Mendel's work. *Bioscene*, 27(2), pp. 13–24.
Moore, D. (2003). *The Dependent Gene: The Fallacy of Nature Vs. Nurture*. New York: Holt.
Moore, D. (2015). *The Developing Genome: An Introduction to Behavioural Epigenetics*. Oxford: OUP.
Morange, M. (2002). The relations between genetics and epigenetics. *Annals of the New York Academy of Sciences*, 981(1), pp. 50–60.
Morange, M. (2006). Post-genomics, between reduction and emergence. *Synthese*, 151(3), pp. 355–360.
Morange, M. (2008). What history tells us XIII. Fifty years of the Central Dogma, *J. Biosci*. 33(2), June 2008, pp. 171–175.
Morgan, T. (1903). *Evolution and Adaptation*. New York: Macmillan & Co.
Morgan, T. (1917). The theory of the gene. *The American Naturalist*, 51, pp. 513–544.
Morgan, T., Sturtevant, A., Muller, H., and Bridges, C. (1915). *The Mechanism of Mendelian Heredity*. New York: Henry Holt.
Morton, P. (1984). *The Vital Science: Biology and the Literary Imagination*. London: Routledge.
Moss, L. (2003). *What Genes Can't Do*. Cambridge, Mass.: MIT Press.
Mossner, N. (2011). Thought styles and paradigms – a comparative study of Ludwik Fleck and Thomas S. Kuhn. *Studies in History and Philosophy of Science*, 42(2), pp. 362–371.
Mottier, V. (2008). Eugenics, politics and the state: social democracy and the Swiss "gardening state". *Studies in History and Philosophy of Science Part C: Studies in History and Philosophy of Biological and Biomedical Sciences*, 39(2), pp. 263–269.
Mottier, V. (2010). Eugenics and the state: policymaking in comparative perspective. In: Bashford, A. and Levine, P., eds. *The Oxford Handbook of the History of Eugenics*. New York: Oxford University Press, pp. 134–153.
Moynihan, D. (1965). *The Negro Family: The Case for National Action*. Washington, D.C.: Office of Policy Planning and Research, U.S. Department of Labor.
Mukerji, C. (1989). *A Fragile Power: Scientists and the State*. Princeton, NJ: Princeton University Press.
Mulkay, M. (1979). *Science and the Sociology of Knowledge*. London: Allen & Unwin.
Müller, G. B. (2010). Epigenetic innovation. In: Pigliucci, M. and Müller, G. B., eds. *Evolution – The Extended Synthesis*. Cambridge, Mass: MIT Press, pp. 307–331.

Muller, H. (1933). The dominance of economics over eugenics. *Scientific Monthly*, 37(1), pp. 40–47.
Muller, H. (1936). *Out of the Night: A Biologist's View of the Future*. London: V. Gollancz, Ltd.
Müller-Hill, B. (1988). *Murderous Science: Elimination by Scientific Selection of Jews, Gypsies, and Others, Germany 1933–1945*. Oxford, UK: Oxford University Press.
Müller-Hill, B. Human genetics and the mass murder of Jews, Gypsies, and others. In: Berenbaum, M. and Peck, A. (1998). *The Holocaust And History: The Known, The Unknown, The Disputed, And The Reexamined*. Bloomington, IN: Indiana University Press, pp. 103–114.
Müller-Wille, S. (2007a). Heredity: the formation of an epistemic space. In: Müller-Wille, S. and Rheinberger, H., eds. *Heredity Produced: At the Crossroads of Biology, Politics, and Culture, 1500–1870*. Cambridge, Mass.: MIT Press, pp. 3–34.
Müller-Wille, S. (2007b). Hybrids, pure cultures, and pure lines: from nineteenth-century biology to twentieth-century genetics, *Studies in History and Philosophy of the Biological and Biomedical Sciences*, 38 (4), pp. 796–806.
Müller-Wille, S. and Rheinberger, H. (2012). *A Cultural History of Heredity*. Chicago: University of Chicago Press.
Muller, H. J. (1950). Our load of mutations. *Am J Hum Genet*. Jun; 2(2), pp. 111–176.
Nägeli, C. (1884). *Mechanisch-physiologische theorie der abstammungslehre*. Munich: Druck und Verlag von R. Oldenbourg.
Nanney, D. (1958). Epigenetic control systems. *Proc. Natl. Acad. Sci. USA*, 44, pp. 712–717.
Nelkin, D. (1975). The political impact of technical expertise. *Social Studies of Science*, 5, pp. 35–54.
Niewöhner, J. (2011). Epigenetics: embedded bodies and the molecularisation of biography and milieu. *BioSocieties*, 6(3), pp. 279–298.
Noble, D. (2011). Neo-Darwinism, the Modern Synthesis and selfish genes: are they of use in physiology? *J Physiol*. 589(Pt 5), pp. 1007–1015.
Noble, G. (1926). Kammerer's alytes. *Nature*, 118(2962), pp. 209–211.
Norton, B. (1983). Fisher's entrance into evolutionary science: the role of eugenics. In: Grene, M., ed. *Dimensions of Darwinism: Themes and Counterthemes in Twentieth-Century Evolutionary Theory*. Cambridge: Cambridge University Press, pp. 9–29.
Novak, J. (2008). *Alfred Russel Wallace and August Weismann's Evolution: A Story Written on Butterfly's Wings*: PhD Dissertation (Princeton University).
Novas, C. and Rose, N. (2005). Biological citizenship. In: Ong, A. and Collier, S., eds. *Global Assemblages: Technology, Politics and Ethics as Anthropological Problems*. Oxford: Blackwell, pp. 439–463.
NY Times. (1923), unsigned, "Biologist to Tell how Species Alter; Dr. Kammerer", Darwin's Suecessor, "Arrives From Vienna for Lectures", 28 November.
Nye, Robert A. (1984). *Crime, Madness and Politics: The Medical Concept of National Decline*. Princeton: Princeton University Press.
O'Hara, R. J. (1997). Population thinking and tree thinking in systematics. *Zoologica Scripta*, 26(4), pp. 323–329.
Odling-Smee, F., Laland, K., and Feldman, M. (2003). *Niche Construction: The Neglected Process in Evolution*. Princeton, NJ: Princeton University Press.

Olby, R. (1966). *Origins of Mendelism*. London: Constable. 2nd ed. (1985), Chicago: University of Chicago Press.
Olby, R. (1970). Francis Crick, DNA, and the Central Dogma, *Daedalus*, 99, pp. 938, 987.
Olby, R. (1974). *The Path to the Double Helix: The Discovery of DNA*. Seattle: University of Washington Press.
Olby, R. (1979). Mendel no Mendelian? *History of Science*, 17, pp. 53–72.
Orel, V. and Hartl, D. L. (1994). Controversies in the interpretation of Mendel's discovery. *History and Philosophy of the Life Sciences*, 16(3), pp. 423–464.
Osborn, F. (1940). *Preface to Eugenics*. New York: Harper and Brothers.
Oyama, S. (2000a). *The Ontogeny of Information*. Durham, NC: Duke University Press.
Oyama, S. (2000b). *Evolution's Eye. A Systems View of the Biology-Culture Divide*, Durham and London: Duke UP.
Oyama, S., Griffiths, P., and Gray, R., eds. (2001). *Cycles of Contingency: Developmental Systems and Evolution*. Cambridge, Mass.: MIT Press.
Packard A. S. (1893). On the Inheritance of Acquired Characters in Animals with a Complete Metamorphosis. *Proceedings of the American Academy of Arts and Sciences*, 29, pp. 331–370.
Packard, A. S. (1901). *Lamarck, the Founder of Evolution: His Life and Work*. London and Bombay: Longmans, Green and Co.
Painter, R., Osmond, C., Gluckman, P., Hanson, M., Phillips, D., and Roseboom, T. (2008). Transgenerational effects of prenatal exposure to the Dutch famine on neonatal adiposity and health in later life. *BJOG: An International Journal of Obstetrics and Gynaecology*, 115, pp. 1243–1249.
Panofsky, A. (2015). From behavior genetics to postgenomics. In: Richardson, S. and Stevens, H., eds. *Postgenomics: Perspectives on Biology after the Genome*. Durham, NC: Duke University Press, pp. 5–172.
Parrinder, P. (1997). Eugenics and utopia: sexual selection from Galton to Morris. *Utopian Studies*, 8, pp. 1–12.
Pastore, N, (1949). *The Nature-Nurture Controversy*. New York: King's Crown Press.
Paul, D. (1984). Eugenics and the left. *Journal of the History of Ideas*, 45(4), pp. 567–590.
Paul, D. (1987). "Our Load of Mutations" Revisited. *Journal of the History of Biology*, 20(3), pp. 321–335.
Paul, D. (1995). *Controlling Human Heredity: 1865 to the Present*. Atlantic Highlands, NJ: Humanities Press International.
Paul, D. (1998). *The Politics of Heredity: Essays on Eugenics, Biomedicine, and the Nature-Nurture Debate*. Albany, NY: State University of New York Press.
Paul, D. (2006). Darwin, social Darwinism, and eugenics. In: Hodge, J. and Radick, G., eds. *The Cambridge Companion to Darwin*. 2nd edition. Cambridge: Cambridge University Press, 219–245.
Paul, D. and Day, B. (2008). John Stuart Mill, innate differences, and the regulation of reproduction, *Studies in History and Philosophy of Biological and Medical Sciences*, 39, pp. 222–231.
Paul, D. and Moore, J. (2010). The Darwinian context: evolution and inheritance. In: Bashford A. and Levine P., eds. *The Oxford Handbook of the History of Eugenics*. New York: Oxford University Press, pp. 27–42.

Pauly, P. (1993). Review: the eugenics industry: growth or restructuring? *Journal of the History of Biology*, 26(1), pp. 131–145.
Pearson, K. (1903). On the Inheritance of the Mental and Moral Characters in Man, and its Comparison with the Inheritance of the Physical Characters. *The Journal of the Anthropological Institute of Great Britain and Ireland*, 33, pp. 179–237.
Pearson, K. (1905). *National Life from the Standpoint of Science*. London: Black.
Pearson, K. (1914, 1924, 1930). *The Life, Letters, and Labours of Francis Galton*. Cambridge: Cambridge University Press.
Pembrey, M. E. (2002). Time to take epigenetic inheritance seriously. *European Journal of Human Genetics*, 10, pp. 669–671.
Pembrey, M., Bygren, L., Kaati, G., Edvinsson, S., Northstone, K., Sjöström, M. and Golding, J. (2006). Sex-specific, male-line transgenerational responses in humans. *Eur J Hum Genet*, 14(2), pp. 159–166.
Pennisi, E. (2008). Modernizing the modern synthesis. *Science*, 321(5886), pp. 196–197.
Pennisi, E. (2009). The case of the midwife toad: fraud or epigenetics?. *Science*, 325(5945), pp. 1194–1195.
Pennisi, E. (2012). ENCODE project writes eulogy for junk DNA. *Science*, 337(6099), pp. 1159–1161.
Pick, D. (1989). *Faces of Degeneration A European Disorder*, c. 1848–c. 1918. Cambridge, UK: Cambridge University Press.
Pickens, D. (1968). *Eugenics and the Progressives*. Nashville, Tennessee: Vanderbilt University Press.
Pickersgill, M., Niewöhner, J., Müller, R., Martin, P. and Cunningham-Burley, S. (2013). Mapping the new molecular landscape: social dimensions of epigenetics. *New Genetics and Society*, 32(4), pp. 429–447.
Pickersgill, M. (2016). Epistemic Modesty, Ostentatiousness and the Uncertainties of Epigenetics: On the Knowledge Machinery of (Social) Science: In: Meloni, M., Williams, S., Martin, P., eds. *Biosocial Matters: Rethinking Sociology-Biology Relations in the Twenty-First Century*. Cambridge, UK: Wiley-Blackwell.
Piersma, T., and van Gils, J. (2011). *The Flexible Phenotype: A Body-Centred Integration of Ecology, Physiology, and Behaviour*. Oxford: Oxford University Press.
Pigliucci, M. (2001). *Phenotypic Plasticity: Beyond Nature and Nurture*. Baltimore: Johns Hopkins University Press.
Pigliucci, M. (2009). An extended synthesis for evolutionary biology. *Annals of the New York Academy of Sciences*, 1168(1), pp. 218–228.
Pigliucci, M. and Müller, G., eds. (2010). *Evolution: The Extended Synthesis*. Cambridge, Mass.: MIT Press.
Pink, R. C. et al. (2011). Pseudogenes: pseudo-functional or key regulators in health and disease? *RNA*, 17(5), pp. 792–798.
Pinker, S. (2002). *The Blank Slate*, New York: Penguin Books.
Plato. (1987). *Republic*, trans. London: Desmond Lee, Penguin.
Plutynski, A. (2006). What was Fisher's fundamental theorem of natural selection and what was it for? *Studies in History and Philosophy of Biological and Biomedical Sciences*, 37, pp. 59–82.
Polanyi, M. (1958). *Personal Knowledge*. London: Routledge and Kegan Paul.
Polanyi, M. (1962). The republic of science: its political and economic theory. *Minerva*, 1(1), pp. 54–73.

Porter, T. (1988). *The Rise of Statistical Thinking: 1820–1900*. Princeton, NJ: Princeton University Press.
Portin, P. (2002). Historical development of the concept of the gene. *J. Med. Philos.*, 27, pp. 257–286.
Portin, P. (2009). The elusive concept of the gene. *Hereditas*, 146(3), pp. 112–117.
Powers J. (2013). Finding Ernst Mayr's Plato. *Studies in History and Philosophy of Science Part C*, 44(4), pp. 714–723.
Prainsack, B. 2015 Is personalized medicine different? (Reinscription: the sequel). A response to Troy Duster. *The British Journal of Sociology*, 66(1), pp. 28–35.
Proctor, R. (1988). *Racial Hygiene: Medicine Under the Nazis*. Cambridge, Mass: Harvard University Press.
Proctor, R. (1999). *The Nazi War on Cancer*. Princeton, NJ: Princeton University Press.
Proctor, R. (2003). Three roots of human recency. *Current Anthropology*, 44, 213–239.
Provine, W. (1986). Genetics and race. *American Zoologist*, 26, pp. 857–887.
Provine, W. (2001). *The Origins of Theoretical Population Genetics*. 2nd edition. Chicago: University of Chicago Press.
Ptashne, M. (2007). On the use of the word "epigenetics." *Curr. Biol.*, 17, pp. R233–R236.
Puri, D. Jyotsna Dhawan & Rakesh K. Mishra. (2010). The paternal hidden agenda: Epigenetic inheritance through sperm chromatin, *Epigenetics*, 5(5), pp. 386–391.
Quinlan, S. (2010). Heredity, reproduction, and perfectibility in revolutionary and Napoleonic France, 1789–1815. *Endeavour*, 34(4), pp. 142–150.
Rabinow, P. and Rose, N. (2006). Biopower today. *BioSocieties*, 1, pp. 195–217.
Rainger, R. (1980). The Henry Fairfield Osborn Papers at the American Museum of Natural History. *The Mendel newsletter; archival resources for the history of genetics & allied sciences*, 18, pp. 8–13.
Rainger, R. (1991). *An Agenda for Antiquity: Henry Fairfield Osborn and Vertebrate Paleontology at the American Museum of Natural History, 1890–1935*. Tuscaloose, AB: University of Alabama.
Rando, O., (2012). Daddy issues: paternal effects on phenotype, *Cell*, 151, pp. 702–708.
Rando, O., and Verstrepen, K. (2007). Timescales of genetic and epigenetic inheritance. *Cell*, 128, pp. 655–668.
Rassoulzadegan, M., Grandjean, V., Gounon, P., Vincent, S., Gillot, I., and Cuzin, F. (2006). RNA-mediated non-mendelian inheritance of an epigenetic change in the mouse. *Nature*, 441, pp. 469–474.
Ravetz, J. (1979). *Scientific knowledge and its social problems*. Oxford: Oxford Univ. Press.
Reardon, J. (2005). *Race to the Finish: Identity and Governance in an Age of Genomics*. Princeton, NJ: Princeton University Press.
Reid, A. (1902). *Alcoholism: A Study in Heredity*. London: T.F. Unwin.
Reid, M. (2011). Behind the "Glasgow effect", *Bulletin of the World Health Organization*, 89(10), pp. 701–776.
Reilly, P. R. (1991). *The Surgical Solution: A History of Involuntary* Sterilization *in the United States*. Baltimore: Johns Hopkins University Press.

Renwick, C. (2012). *British Sociology's Lost Biological Roots: A History of Futures Past.* Houndmills, Basingstoke, Hampshire: Palgrave Macmillan.
Renwick, C. (2016). New bottles for new wine: Julian Huxley, biology and sociology in Britain. In: Meloni, M., Williams, S., Martin, P., eds. *Biosocial Matters: Rethinking Sociology-Biology Relations in the Twenty-First Century.* Oxford: Wiley-Blackwell.
Reynolds D. (2001). *One World Divisible: A Global History Since 1945.* New York: Norton and C.
Reynolds, G. (2014). How Exercise Changes Our DNA. New York Times, December 17.
Rheinberger, H.-J. (1997). *Toward a History of Epistemic Things. Synthesizing Proteins in the Test Tube.* Stanford: Stanford University Press.
Rheinberger, H.-J. (2003). Gene Concepts. Fragments from the perspective of molecular biology. In: Beurton, P., Falk, R., Rheinberger, H. J., eds. *The Concept of the Gene in Development and Evolution. Historical and Epistemological Perspectives.* Cambridge: Cambridge University Press, pp. 219–239.
Rheinberger, H.-J. (2005). Reassessing the historical epistemology of Georges Canguilhem. In: Gutting, G., ed. *Continental Philosophy of Science.* Malden, MA, USA: Blackwell Publishing Ltd.
Rheinberger, H.-J. (2010). *On Historicizing Epistemology.* Stanford, CA: Stanford University Press.
Rheinberger, H.-J. (2013). Heredity in the twentieth century: some epistemological considerations. *Public Culture,* 25(371), pp. 477–493.
Rheinberger, H.-J., Müller-Wille, S. and Meunier, R. (2015). Gene. [online] In: Zalta, E., ed., *Stanford Encyclopedia of Philosophy.* Available at http://plato.stanford.edu/archives/spr2015/entries/gene/. Accessed 24 June, 2015.
Richards, E. (2006). Inherited epigenetic variation: revisiting soft inheritance. *Nature Reviews Genetics,* 7, pp. 395–401.
Richards, R. (1987). *Darwin and the Emergence of Evolutionary Theories of Mind and Behavior.* Chicago: University of Chicago Press.
Richards, R. (1992). *The Meaning of Evolution: The Morphological Construction and Ideological Reconstruction of Darwin's Theory.* Chicago: University of Chicago Press.
Richards, M. (2013). Review of The Oxford handbook of the history of eugenics. *New Genetics and Society,* 32(3), pp. 322–328.
Richardson, S. (2012). *Sex Itself: The Search for Male and Female in the Human Genome.* Chicago: Chicago UP.
Richardson, S. (2015). Maternal bodies in the postgenomic order: gender and the explanatory landscape of epigenetics. In: Richardson, S. and Stevens, H., eds. *Postgenomics: Perspectives on Biology after the Genome.* Durham, NC: Duke University Press, pp. 210–231.
Richardson, S., Daniels, C., Gillman, M., Golden, J., Kukla, R., Kuzawa, C., and Rich-Edwards, J. (2014). Society: don't blame the mothers. *Nature,* 512, pp. 131–132.
Rife, D. (1945). *The Dice of Destiny.* Columbus, OH: Long's College Book Co.
Robert, J. (2004). *Embriology, Epigenesis and Evolution: Taking Development Seriously.* New York: Cambridge University Press.
Robinson, G. (1979). *A Prelude to Genetics: Theories of a Material Substance of Heredity, Darwin to Weismann.* Lawrence, KS: Coronado Press.

Rockell F. (1912). The Last of the Great Victorians: Special Interview with Dr. Alfred Russel Wallace. *The Millgate Monthly*, August accessed at http://people.wku.edu/charles.smith/wallace/S750.htm.
Rodwell, G. (1997). Dr Caleb Williams Saleeby: the complete eugenicist. *History of Education*, 26(1), pp. 23–40.
Roll-Hansen, N. (2004). *The Lysenko Effect: The Politics of Science*. Amherst, NY: Humanity Books.
Roll-Hansen, N. (2005). The Lysenko effect: undermining the autonomy of science. *Endeavour*, 29(4), pp. 143–147.
Roll-Hansen, N. (2009). Sources of Wilhelm Johannsen's genotype theory. *Journal of the History of Biology*, 42(3), pp. 457–493.
Roll-Hansen, N. (2011). Lessons from the history of science. *Studies in History and Philosophy of Science Part A*, 42(3), pp. 462–466.
Romanes, G. (1883). *Mental Evolution in Animals, with a Posthumous Essay on Instinct by Charles Darwin*. London: Kegan Paul, Trench & Co.
Romanes, G. (1888). *Mental Evolution in Man*. London: Kegan Paul, Trench & Co.
Romanes, G. (1889). Mr. Wallace on Darwinism. *Contemporary Review*, 56, pp. 244–258.
Romanes, G. (1895). *Darwin, and after Darwin, vol. 2*. Chicago: The Open Court Publishing Company.
Romanes, G. (1899). An *Examination* of *Weismannism*. Chicago: Open Court (2nd edition).
Rose, N. (1979). The Psychological Complex: Mental Measurement and Social Administration, *Ideology and Consciousness*, 5, pp. 5–68.
Rose, N. (2007). *The Politics of Life Itself: Biomedicine, Power, and Subjectivity in the Twenty-First Century*. Princeton, NJ: Princeton University Press.
Rose, M. and Oakley, T. (2007). The new biology: beyond the Modern Synthesis. *Biol Direct*, 2(1), p. 30.
Rosenberg, C. (1967). Factors in the development of genetics in the United States: some suggestions. *Journal of the History of Medicine and Allied Sciences*, 22(1), pp. 27–46.
Rosenberg, C. (1974). The bitter fruit: heredity, disease and social thought in nineteenth century America. *Persp. Am. Hist.*, 1, pp. 189–235.
Ross, E. (1901). The causes of racial superiority. *Annals of the American Academy of Political and Social Science*, 18, pp. 67–89.
Rothstein, M., Cai, Y., and Marchant, G. (2009). The ghost in our genes: legal and ethical implications of epigenetics. *Health Matrix*, 19, pp. 1–62.
Rubin, B. (2009). Changing brains: the emergence of the field of adult neurogenesis. *BioSocieties*, 4, pp. 407–424.
Saavedra-Rodriguez, L. and Feig, L. (2013). Chronic social instability induces anxiety and defective social interactions across generations. *Biol. Psychiatry*, 73, pp. 44–53.
Saleeby, C. (1906). *Ethics*. London: publisher unknown.
Saleeby, C. (1909). *Parenthood and race culture: an outline of eugenics*. New York: Moffat, Yard, and Company.
Saleeby, C. (1910a). Racial poisons II. Alcohol. *Eugen Rev*, 2(1), pp. 30–52.
Saleeby, C. (1910b). The methods of eugenics. *The Sociological Review*, a3(4), pp. 277–286.
Saleeby, C. (1910c). *Professor Karl Pearson on Alcoholism and Offspring*.

Saleeby, C. (1911). *The Methods of Race-Regeneration*. New York: Moffat, Yard and Company.
Saleeby, C. (1914). *The progress of eugenics*. London: Cassell and Company, Ltd.
Sampson, R. (2003). The neighborhood context of well-being. *Perspectives in Biology and Medicine*, 46, pp. S53–S64.
Sapp, J. (1983). The struggle for authority in the field of heredity, 1900–1932: new perspectives on the rise of genetics. *Journal of the History of Biology*, 16, pp. 311–342.
Sapp, J. (1987). *Beyond the Gene: Cytoplasmic Inheritance and the Struggle for Authority in Genetics*. New York: Oxford University Press.
Sapp, J. (1990). *Where the Truth Lies: Franz Moewus and the Origin of Molecular Biology*. Cambridge, UK: Cambridge UP.
Sapp, J. (2003). *Genesis: The Evolution of Biology*. Oxford: Oxford UP.
Schiller, F. (1930). Eugenics as a moral ideal: the beginning of a progressive reform. *Eugenics Review*, 22(2), pp. 103–109.
Schlichting, C. and Pigliucci, M. (1998). *Phenotypic Evolution: A Reaction Norm Perspective*. Sunderland, Mass.: Sinauer.
Schneider, W. (1990). The eugenics movement in France, 1890–1940. Adams, M. ed. *The Wellborn Science: Eugenics in Germany, France, Brazil, and Russia*. New York: Oxford University Press, pp. 69–109.
Scott, J. (1998). *Seeing Like a State: How Certain Schemes to Improve the Human Condition Have Failed*. New Haven: Yale University Press.
Searle G. R. (1971). *The Quest for National Efficiency: A Study in British Politics and Political Thought, 1899–1914*. Berkeley, CA: University of California Press.
Searle, G. ed. (1976). *Eugenics and Politics in Britain, 1900–1914*. Leyden: Noordhoff International Publishing.
Searle, G. (1979). Eugenics and politics in Britain in the 1930s. *Annals of Science*, 36(2), pp. 159–169.
Searle, G. (1981). Eugenics and Class. In: Webster, C, ed. *Biology, Medicine and Society 1840–1940*. Cambridge, UK: Cambridge University Press.
Segerstråle, U. (2000), *Defenders of the Truth*, Oxford University Press, Oxford.
Segerstråle, U. (2013). *Nature's Oracle. The Life and Work of W.D. Hamilton*. Oxford: Oxford University Press.
Selcer, P. (2012). Beyond the cephalic index: negotiating politics to produce UNESCO's scientific statements on race. *Current Anthropology*, 53(S5), pp. S173–S184.
Selden S. (1999). *Inheriting Shame: The Story ofEugenics and Racism in America*. New York: Teachers College Press.
Shapin, S. (1979). Homo Phrenologicus: anthropological perspectives on an historical problem. In: Barnes, Barry and Shapin, Steven, eds. *Natural Order: Historical Studies in Scientific Culture*. Beverly Hills, CA: Sage, pp. 41–71.
Shapin, S. (1992). Discipline and bounding: the history and sociology of science as seen through the externalism-internalism debate. *History of Science*, 30, pp. 333–369.
Shapin, S. (1995). Here and everywhere – sociology of scientific knowledge. *Annual Review of Sociology*, 21, pp. 289–321.
Shapin, S. (2010). *Never Pure: Historical Studies of Science as if It Was Produced by People with Bodies, Situated in Time, Space, Culture, and Society, and Struggling for Credibility and Authority*, Baltimore: Johns Hopkins.

Shapin, S. and Schaffer, S. (1985). *Leviathan and the Air-Pump: Hobbes, Boyle, and the Experimental Life*. Princeton NJ: Princeton University Press.
Shapiro J, (2009). Revisiting the central dogma in the 21st century. Natural Genetic Engineering and Natural Genome Editing. *Ann. N.Y. Acad. Sci.*, 1178, pp. 6–28.
Sharkey, P. (2008). The intergenerational transmission of context. *American Journal of Sociology*, 113, pp. 931–969.
Sharkey, P. (2014). *Stuck in Place: Urban Neighborhoods and the End of Progress toward Racial Equality*. Chicago: University of Chicago Press.
Shavit, A. (2004). Shifting values partly explain the debate over group selection. *Stud. Hist. Phil. Biol. & Biomed. Sci.* Part C, 35(4), pp. 697–720.
Shaw, M., Tunstall, H., and Davey Smith, G. (2003). Seeing social position: visualizing class in life and death. *International Journal of Epidemiology*, 32, pp. 332–335.
Shelden, S. (1999). *Inheriting Shame: The Story of Eugenics and Racism in America*. New York: Teacher College Press.
Shermer, M. (2001). The Pinker Instinct. Altadena, CA: *Skeptics Society & Skeptic Magazine*.
Simpson, G. G. (1944). *Tempo and Mode in Evolution*. New York: Columbia University Press.
Simpson, G. G. (1949). *The Meaning of Evolution: A Study of the History of Life and of Its Significance for Man*. New Haven: Yale University Press.
Simpson, G. G. (1964). *This View of Life: The World of an Evolutionist*. New York: Harcourt, Brace & World.
Simpson, G. G. (1969). *Biology and Man*. New York: Harcourt, Brace & World.
Simpson, G. G. (1941). The Role of the Individual in Evolution, *Journal of the Washington Academy of Sciences*, 31, pp. 1–20.
Sinnott, E. W. (1945–1946). The biological basis of democracy. *The Yale Review*, 35, pp. 61–73.
Slack, J. (2002). Conrad Hal Waddington: the last Renaissance biologist? *Nature Reviews Genetics*, 3, pp. 889–895.
Sluga, G. (2010). UNESCO and the (one) world of Julian Huxley. *Journal of World History*, 21, pp. 1–18.
Smocovitis, V. (1992). Unifying biology: The evolutionary synthesis and evolutionary biology. *J Hist Biol*, 25(1), pp. 1–65.
Smocovitis, V. (1994). Organizing evolution: founding the society for the study of evolution (1939–1950). *J Hist Biol*, 27(2), pp. 241–309.
Smocovitis, V. (1996). *Unifying Biology: The Evolutionary Synthesis and Evolutionary Biology*. Princeton: Princeton University Press.
Smocovitis, V. (1999). The 1959 Darwin centennial celebration in America. *Osiris*, 14(1), pp. 274–323.
Smocovitis, V. (2009). The unifying vision: Julian Huxley, the evolutionary synthesis and evolutionary humanism. In: Somsen, Geert and Kamminga, Harmke, eds. *Pursuing The Unity of Science: Ideology and Scientific Practice between the Great War and the Cold War*. London: Ashgate.
Smocovitis, V. (2012). Humanizing evolution: anthropology, the evolutionary synthesis, and the prehistory of biological anthropology, 1927–1962. *Current Anthropology*, 53 (S5), pp. S108–S125.
Smythies, J., Edelstein, L., Ramachandran, V. (2014). Molecular mechanisms for the inheritance of acquired characteristics–exosomes, microRNA shuttling, fear and stress: Lamarck resurrected?, *Front Genet*, 5, p. 133.

Sober, E., (1980), Evolution, population thinking and essentialism, *Philosophy of Science*, 47, pp. 350–383.

Solovey, M. and Cravens, H., eds. (2012). *Cold War Science: Knowledge Production, Liberal Democracy, and Human Nature*. New York: Palgrave Macmillan.

Soloway, R. (1990). *Demography and Degeneration: Eugenics and the Declining Birthrate in Twentieth-Century Britain*. Chapel Hill, NC: University of North Carolina Press.

Sonneborn, T. (1949). Beyond the gene. *American Scientist*, 37, pp. 33–59.

Sparks, C. S., Jantz, R. L. (2003). Changing times, changing faces: Franz Boas's immigrant study in modern perspective. *American Anthropologist*, 105(2), p. 333.

Spencer, H. (1893a). The inadequacy of "natural selection," *Contemporary Review.*, 63, pp. 152–166, 439–456.

Spencer, H. (1893b). A Rejoinder to Prof. Weismann, *Contemporary Review*, 64, pp. 893–912.

Spencer, H. (1894). Weismann once more. *Contemporary Review*, 66, pp. 592–608.

Stanford, K. (2006). *Exceeding Our Grasp: Science, History, and the Problem of Unconceived Alternatives*. New York: Oxford University Press.

Staum, M. (2011). *Nature and Nurture in French Social Sciences, 1859–1914 and beyond*. Montreal: McGill-Queens University Press.

Stebbins G. L. (1950). *Variation and Evolution in Plants*. New York: Columbia University Press.

Steggerda, M. and Charles Benedict, D. (1929). *Race Crossing in Jamaica*. Carnegie Institution of Washington Publication, p. 395.

Steinfels, P. (1986). Biomedical ethics and the shadow of Nazism: A conference on the proper use of Nazi analogy in ethical debate. *The Hastings Center Report*, 6, pp. 1–19.

Stepan, N. (1982). *The Idea of Race in Science: Great Britain, 1800–1960*. Hamden, Conn.: Archon Books.

Stepan, N. (1991). *The Hour of Eugenics: Race, Gender, and Nation in Latin America*. Ithaca, NY: Cornell University Press.

Stern, C. (1950). Concluding Remarks of the Chairman. In: Demerec, M., ed. *Origin and Evolution of Man*. Cold Spring Harbor Symposia on Quantitative Biology, vol. 15. Cold Spring Harbor, NY: Biological Laboratory.

Stern, C. M. (2005). Sterilized in the Name of Public Health: Race, Immigration, and Reproductive Control in Modern California, *Am J Public Health*, 95(7), pp. 1128–1138.

Stevens, H. and Richardson, S. (2015). Approaching postgenomics. In: Richardson, S. and Stevens, H., eds. *Postgenomics: Perspectives on Biology after the Genome*. Durham, NC: Duke University Press, pp. 232–241.

Stillwell, D. (2012). Eugenics visualized: the exhibit of the Third International Congress of Eugenics, 1932. *Bulletin of the History of Medicine*, 86(2), 206–236.

Stites, R. (1989). *Revolutionary Dreams: Utopian Vision and Experimental Life in the Russian Revolution*. New York. Oxford. Oxford University Press.

Stocking, G. (1968). *Race, Culture, and Evolution: Essays in the History of Anthropology*. Chicago: University of Chicago Press.

Stocking, G. (1987). *Victorian Anthropology*. New York: The Free Press.

Stocking, G. (2001). *Delimiting Anthropology: Occasional Inquiries and Reflection*. Madison: University of Wisconsin Press.

Stoczkowski W. (2009). UNESCO's doctrine of human diversity: A secular soteriology? *Anthropology Today*, 25(3), pp. 7–11.
Stone, D. (2001). Race in British eugenics. *European History Quarterly* (July), 31(3), pp. 397–425.
Stone, D. (2002). *Breeding Superman: Nietzsche, Race and* Eugenics *in Edwardian and Interwar Britain*. Liverpool: Liverpool University Press.
Stotz, K. (2006). With genes like that, who needs an environment? Postgenomics' argument for the ontology of information. *Philos. Sci.*, 73, pp. 905–917.
Stotz, K. (2008). The ingredients for a postgenomic synthesis of nature and nurture. *Philos. Psychol.*, 21, pp. 359–381.
Strasser, B. (2006). A world in one dimension: Linus Pauling, Francis Crick and the Central Dogma of Molecular Biology. *Hist. Phil. Life Sci.*, 28, pp. 491–512.
Strasser, B. (2012). Data-driven sciences: from wonder cabinets to electronic databases, *Studies in History and Philosophy of Biological and Biomedical Sciences*, 43, pp. 85–87.
Sullivan, S. (2013). Inheriting racist disparities in health: epigenetics and the transgenerational effects of white racism. *Critical Philosophy of Race*, 1(2), pp. 190–218.
Sultan, S. (2007). Development in context: the timely emergence of eco-devo. *Trends in Ecology & Evolution*, 22(11), pp. 575–582.
Sunder Rajan, K. (2006). *Biocapital: The Constitution of Postgenomic Life*. Durham, NC: Duke University Press.
Sunder Rajan, K. and Leonelli, S. (2013). Introduction: biomedical trans-actions, postgenomics and knowledge/value. *Public Culture*, 25(3), pp. 463–475.
Szyf, M. (2001). Towards a pharmacology of DNA methylation. *Trends in Pharmacological Sciences*, 22(7), pp. 350–354.
Szyf, M. (2009a). The early life environment and the epigenome. *Biochimica et Biophysica Acta (BBA) – General Subjects*, 1790(9), pp. 878–885.
Szyf, M. (2009b). Implications of a life-long dynamic epigenome. *Epigenomics*, 1, pp. 9–12.
Szyf, M. (2013). Lamarck revisited: epigenetic inheritance of ancestral odor fear conditioning. *Nature Neuroscience*, 17(1), pp. 2–4.
Szyf, M., and Bick, J. (2013). DNA methylation: a mechanism for embedding early life experiences in the genome. *Child Development*, 84(1), pp. 49–57.
Tabery, J. G. (2008). R. A. Fisher, Lancelot Hogben, and the Origin(s) of Genotype–Environment Interaction, *Journal of the History of Biology*, 41, pp. 717–761.
Taguieff, P. A. (1991). sur l'eugénisme: du fantasme au débat. *Povouir*, 56, pp. 23–64.
Tal, O., Kisdi, E. and Jablonka, E. (2010). Epigenetic contribution to covariance between relatives, *Genetics*, 184(4), pp. 1037–1050.
TallBear, K. (2013). *Native American DNA: Tribal Belonging and the False Promise of Genetic Science*. Minneapolis, MN: University of Minnesota Press.
Tax S., ed. (1960). *Evolution after Darwin*. 3 vols. Chicago: University of Chicago Press, Chicago.
Taylor 1988 Technocratic Optimism, H. T. Odum, and the Partial Transformation of Ecological. *Journal of the History of Biology*, 21(2), pp. 213–244.
Thacker, E. (2005). *The Global Genome: Biotechnology, Politics, and Culture*. Cambdridge, Mass: MIT Press.

Thayer, Z. and Kuzawa, C. (2011). Biological memories of past environments: Epigenetic pathways to health disparities. *Epigenetics*, 6(7), pp. 798–803.
Tolwinski, K. (2013). A new genetics or an epiphenomenon? Variations in the discourse of epigenetics researchers. *New Genetics and Society*, 32(4), pp. 366–384.
Trivers, R. (1971). The Evolution of Reciprocal Altruism, *Quarterly Review of Biology*, 46, pp. 35–57.
Turda, M. (2010). *Modernism and Eugenics*. New York: Palgrave Macmillan.
Turda, M. and Weindling, P. eds. (2007). *Blood and Homeland Eugenics and Racial Nationalism in Central and Southeast Europe, 1900–1940*, Budapest, New York: Central European University Press.
Turner, B. S. (2009). The sociology of the body. In: *Turner*, B. S., ed. *The New Blackwell Companion to Social Theory*. Oxford: Blackwell, pp. 513–532.
Turner, F. (1974). *Between Science and Religion: Reaction to Scientific Naturalism in Late Victorian England*. New Haven, CT: Yale University Press.
Turner, S. (2007). *Political Epistemology, Experts and the Aggregation of. Knowledge*, Accessed at http://spontaneousgenerations.library.utoronto.ca/index.php/SpontaneousGenerations/article/view/2970.
Turner, S. (2007). A life in the first half-century of sociology: Charles Ellwood and the division of sociology. In: Calhoun, C. ed. *Sociology in America: A History*. Chicago and London: University of Chicago Press.
Uller, T. (2013). Non-genetic inheritance and evolution. In: Kampourakais, K., ed. *Philosophy of Biology: A Companion for Educators*. Springer: Verlag.
UNESCO (1969). *Four Statements on the Race Question*. Paris: Unesco.
Vargas, A. (2009). Did Paul Kammerer discover epigenetic inheritance? A modern look at the controversial midwife toad experiments. *Journal of Experimental Zoology Part B: Molecular and Developmental Evolution*, 312B(7), pp. 667–678.
Vineis, P., Stringhini, S., and Porta, M. (2014). The environmental roots of non-communicable diseases (NCDs) and the epigenetic impacts of globalization. *Environmental Research*, 133, pp. 424–430.
Waddington, C. (1957). *The Strategy of the Genes*, London: Allen and Unwin.
Waddington, C. (1968). *Towards a Theoretical Biology*, vol. 1. Edinburgh: Edinburgh University Press.
Waddington, C. (2012, reprinted). The epigenotype. *Int. J. Epidemiol.*, 41, pp. 10–13.
Wade, N. (2009). From one genome, many types of cells. But how? *New York Times*, February 23.
Waggoner, M. and Uller, T. (2015). Epigenetic determinism in science and society, *New Genetics and Society*, 34(2), pp. 177–195.
Wagley, C. (1964). Alfred Métraux (1902–1963), *American Anthropologist*, 66(3), pp. 603–613.
Wagner, G. (2009). Paul Kammerer's midwife toads: about the reliability of experiments and our ability to make sense of them. *Journal of Experimental Zoology Part B: Molecular and Developmental Evolution*, 312B(7), pp. 665–666.
Wagner, K., Wagner, N., Ghanbarian, H., Grandjean, V., Gounon, P., Cuzin, F., et. al. (2008). RNA induction and inheritance of epigenetic cardiac hypertrophy in the mouse. *Dev. Cell*, 14, pp. 962–969.
Wallace, A. R. (1892). Human progress: past and future. *Arena*, 5 (26), pp. 145–159.

Wallace A. R. An interview by Frederick *Rockell* printed in The Millgate Monthly of August *1912*, Accessed at http://people.wku.edu/charles.smith/wallace/S750.htm on July 2015.
Wallace B and Dobzhansky T. (1959). *Radiation, Genes, and Man*. New York: Holt.
Waller, J. (2001). Ideas of heredity, reproduction and eugenics in Britain, 1800–1875. *Studies in History and Philosophy of Science Part C: Studies in History and Philosophy of Biological and Biomedical Sciences*, 32(3), pp. 457–489.
Waller, J. (2002). "The illusion of an explanation": the concept of hereditary disease, 1770–1870. *Journal of the History of Medicine and Allied Sciences*, 57(4), pp. 410–448.
Waller, J. (2002). Putting method first: re-appraising the extreme determinism and hard hereditarianism of Sir Francis Galton. *History of Science*, 40(1), pp. 35–62.
Walters, K. (2011). Bodies don't just tell stories, they tell histories: embodiment of historical trauma among American Indians and Alaska Natives. *Du Bois Review: Social Science Research on Race*, 8(1), pp. 179–189.
Ward, L. (1891). *Neo-Darwinism and Neo-Lamarckism*. Washington: Press of Gedney & Roberts.
Washburn, S., ed. (1963). *Classification and Human Evolution*. Chicago: Aldine Pub. Co.
Waterland, R., and Jirtle, R. (2003). Transposable elements: targets for early nutritional effects on epigenetic gene regulation. *Molecular and Cellular Biology*, 23, pp. 5293–5300.
Waters, C. (1987). Introduction: revising our picture of Julian Huxley. In: Waters, C. and Van Helden, A., eds. (1992). *Julian Huxley, Biologist and Statesman of Science*. Houston: Rice University Press, p. 20.
Waters, C. and Van Helden, A., eds. (1992). *Julian Huxley, Biologist and Statesman of Science*. Houston: Rice University Press.
Watson, J. D. and Crick, F. H. C. (1953a). A structure for deoxyribose nucleic acid. *Nature*, 171, pp. 737–738.
Watson, J. D. and Crick, F. H. C. (1953b). Genetical implications of the structure of deoxyribonucleic acid. *Nature*, 171, pp. 964–967.
Weaver, I., Cervoni, N., Champagne, F., D'Alessio, A., Sharma, S., Seckl, J., Dymov, S., Szyf, M. and Meaney, M. (2004). Epigenetic programming by maternal behavior. *Nature Neuroscience*, 7(8), pp. 847–854.
Weidman, N. (2012). An Anthropologist on TV: Ashley Montagu and the Biological Basis of Human Nature, 1945–1960. In*The Social Sciences in the Cold War*, ed. Mark Solovey and Hamilton Cravens, Palgrave, pp. 215–232.
Weindling, P. (1989). *Health, Race and German Politics between National Unification and Nazism: 1870–1945*. Cambridge: Cambridge University Press.
Weindling, P. (2004). *Nazi Medicine and the Nuremberg Trials: From Medical War Crimes to Informed Consent*. Basingstoke, UK: Palgrave Macmillan.
Weindling, P. (2012). Julian Huxley and the continuity of eugenics in twentieth-century Britain. *Journal of Modern European History*, 10(4), pp. 480–499.
Weismann, A. (1883). *Die entstehung der sexualzellen bei den hydromedusen: zugleich ein beitrag zur kenntniss des baues und der lebenserscheinungen dieser gruppe*. Jena: Fischer.
Weismann, A. (1885). *Die continuität des keimplasmas als grundlage einer theorie der vererbung*. Jena: Fischer.

Weismann, A. (1891). *Essays upon Heredity and Kindred Problems*, vol. 1. Schoenland, E. and Shipley, A., eds. Oxford: Oxford University Press.
Weismann, A. (1893a). *The Germ-Plasm: A Theory of Heredity*. London: Charles Scribner's Sons.
Weismann, A. (1893b). The all-sufficiency of natural selection. A reply to Herbert Spencer. Contemp. Rev. 64, pp. 309–338.
Weismann, A. (1904). *The Evolution Theory*, vol. 2.
Weiss, S. (1987). The race hygiene movement in Germany. *Osiris*, 3(1), p. 193.
Weiss, S. (1990). The Race Hygiene Movement in Germany. In: Adams, Mark B, ed. *The Wellborn Science*.
Weiss, S. (2010). *The Nazi Symbiosis: Human Genetics and Politics in the Third Reich*, Chicago: University of Chicago Press.
Weissmann, G. (2010). The Midwife Toad and Alma Mahler: Epigenetics or a Matter of Deception? *The FASEB Journal*, 24 (8), pp. 2591–2595.
Wells, H. G. (1905). *A Modern Utopia*. London: Thomas Nelson and Sons.
Wells, H. G., Huxley, J. and Wells, G. P. (1931). *The Science of Life*. Garden City, N.Y.: Doubleday, Doran & Co.
Wells, J. (2003). Parent-offspring conflict theory, signaling of need, and weight gain in early life. *The Quarterly Review of Biology*, 78(2), pp. 169–202.
Wells, J. (2007a). The thrifty phenotype as an adaptive maternal effect. *Biological Reviews*, 82(1), pp. 143–172.
Wells, J. (2007b). Environmental quality, developmental plasticity and the thrifty phenotype: a review of evolutionary models. *Evol Bionform Online*, 3, pp. 109–120
Wells, J. (2010). Maternal capital and the metabolic ghetto: An evolutionary perspective on the transgenerational basis of health inequalities. *American Journal of Human Biology*, 22(1), pp. 1–17.
Welshman, J. (2013). *Underclass: A History of the Excluded Since 1880*. London: Bloomsbury.
Werskey, G. (1979). *The Visible College: The Collective Biography of British Scientific Socialists of the 1930s*. New York: Holt, Rinehart, and Winston.
West-Eberhard, M. (1992). Genetics, epigenetics, and flexibility: a reply to Crozier. *The American Naturalist*, 139(1), pp. 224–226.
West-Eberhard, M. (2003). *Developmental Plasticity and Evolution*. Oxford: Oxford University Press.
Williams, G. (1966). *Adaptation and Natural Selection*. Princeton, NJ: Princeton University Press.
Williams, G. (1989). A sociobiological expansion of evolution and ethics. In: Huxley, T., Paradis, J. and Williams, G., eds. (1989). *Evolution & ethics*. Princeton, NJ: Princeton University Press, pp. 179–214.
Wilkins, A. S. (2002). Interview with Ernst Mayr. *BioEssays*, 24, pp. 960–973.
Wilson, E. O. (1978). *On Human Nature*. Cambridge, MA: Harvard University Press.
Wilson, P. K. (2007). Erasmus Darwin and the 'noble' disease (Gout): conceptualizing heredity and disease in enlightenment England. In: Müller-Wille, Staffan and Rheinberger, Hans-Jörg, eds. *Heredity Produced*. Cambridge, Mass: MIT Press, pp. 133–154.
Wilson, T. W., and Grim, C. E. (1991). Biohistory of Slavery and Blood Pressure Differences in Blacks Today. A Hypothesis. *Hypertension* 17(1), pp. 122–128.

Wynne, B. C. G. (1976). *Barkla* and the *J-phenomenon*: a case study on the treatment of deviance in physics. *Social Studies of Science*, 6, pp. 307–347.

Winsor M. P. (2006). The creation of the essentialism story: an exercise in metahistory. *History and Philosophy of the Life Sciences*, 28, pp. 149–174.

Winther, R. (2000). Darwin on variation and heredity. *Journal of the History of Biology*, 33(3), pp. 425–455.

Winther, R. (2001). August Weismann on germ-plasm variation. *Journal of the History of Biology*, 34, pp. 517–555.

Witteveen, J. (unpublished). A temporary oversimplication: Mayr, Simpson, Dobzhansky, and the origins of the typology/population dichotomy.

Woiak, J. (1998). Drunkenness, degeneration, and eugenics in Britain, 1900–1914. PhD dissertation, University of Toronto.

Wolpoff, M. H. & Caspari, R. (1997). *Race and Human Evolution: A Fatal Attraction*. New York: Simon and Schuster.

Yadav, S. (2007). The wholeness in suffix *-omics,-omes,* and the word *om*. *J Biomol Tech*, 18(5), p. 227.

Yehuda R. et al. (2005). Transgenerational effects of posttraumatic stress disorder in babies of mothers exposed to the World Trade Center attacks during pregnancy, *J Clin Endocrinol Metab*, 90(7), pp. 4115–4118.

Yehuda R. et al. (2014). Influences of maternal and paternal PTSD on epigenetic regulation of the glucocorticoid receptor gene in Holocaust survivor offspring, *Am J Psychiatry*, 171(8), pp. 872–880.

Young, R. M. (1982). Darwinism is social. In: Kohn, D., ed., The Darwinian Heritage. Princeton University Press, Princeton, NJ, pp. 609–638.

Yudell, M. (2014). *Race Unmasked: Biology and Race in the Twentieth Century*. New York: Columbia University Press.

Zammito, J. H. (2002). *Kant, Herder, and the Birth of Anthropology*. Chicago: University of Chicago Press.

Zammito, J. H. (2004). *A nice derangement of epistemes: Post-positivism in the study of science from Quine to Latour*. University of Chicago Press, Chicago & London.

Zirkle, C. (1946). The early history of the idea of the inheritance of acquired characters and of pangenesis. *Transactions of the American Philosophical Society*, 35(2), p. 91.

Zirkle, C. 1949 *Death of a Science in Russia*. Philadelphia, PA: University of Pennsylvania Press.

Zuckerandl, E. (1962). Perspectives on molecular anthropology. In: Washburn, S., ed. (*Classification and human evolution*. Chicago: Aldine Pub. Co, pp. 243–272.

Index

Printed in the United States
By Bookmasters